Ecological Conservation and Environmental Protection in China, 1978–2018

Ecological Conservation and Environmental Protection in China, 1978–2018

Edited by
Pan Jiahua and Zhuang Guiyang

PETER LANG
New York · Berlin · Bruxelles · Chennai · Lausanne · Oxford

Library of Congress Cataloging-in-Publication Data

Names: Jiahua, Pan, editor. | Guiyang, Zhuang, editor.
Title: Ecological conservation and environmental protection in China, 1978–2018 / edited by Pan Jiahua and Zhuang Guiyang.
Description: New York : Peter Lang Publishing, Inc., 2025. | Includes bibliographical references.
Identifiers: LCCN 2024051541 (print) | LCCN 2024051542 (ebook) | ISBN 9781636678535 (hardback) | ISBN 9781636678542 (ebook) | ISBN 9781636678559 (epub)
Subjects: LCSH: Environmental policy–China. | Nature conservation–China. | Environmental protection–China. | Economic development–Environmental aspects–China. | China–Environmental conditions.
Classification: LCC GE190.C6 E36 2025 (print) | LCC GE190.C6 (ebook) | DDC 333.720951/09045–dc23/eng/20241214
LC record available at https://lccn.loc.gov/2024051541
LC ebook record available at https://lccn.loc.gov/2024051542
DOI 10.3726/b21669

Bibliographic information published by the Deutsche Nationalbibliothek.
The German National Library lists this publication in the German National Bibliography; detailed bibliographic data is available on the Internet at http://dnb.d-nb.de.

Cover design by Peter Lang Group AG

Translated by Li Rong
Proofread by Zhang Haiying

ISBN 9781636678535 (hardback)
ISBN 9781636678542 (ebook)
ISBN 9781636678559 (epub)
DOI 10.3726/b21669

This edition is an authorized translation from the Chinese language edition
Published by arrangement with Social Sciences Academic Press (China)

All rights reserved
The publication of this book is sponsored by the Chinese Fund for the Humanities and Social Sciences

© 2025 Peter Lang Group AG, Lausanne
Published by Peter Lang Publishing Inc., New York, USA
info@peterlang.com - www.peterlang.com

All rights reserved.
All parts of this publication are protected by copyright.
Any utilization outside the strict limits of the copyright law, without the permission of the publisher, is forbidden and liable to prosecution.
This applies in particular to reproductions, translations, microfilming, and storage and processing in electronic retrieval systems.

This publication has been peer reviewed.

Preface

Ecological conservation and environmental protection in China can be traced far back into history. The ancient Chinese philosophical wisdom of the unity of heaven and man, and the 5,000-year farming civilization that has full respect for nature is the very foundation for China's pursuit of environmental health in the past four decades of reform and opening-up. The environmental protection in a strict sense, however, stemmed from the pollutant discharge and pollution management in the industrial age. China has learned from the West when advancing its own industrialization and urbanization. Such learning, be it proactive or passive, means to use the experience from the western hemisphere for the country's own good. This course, undoubtedly, involves the reform of existing institutions and the opening-up to the Western world. China's reform and opening-up since 1978, which has been of large scale, in fast pace, with significant results, long duration and far-reaching influence, is rarely seen in human history of economic development. Moreover, the efforts in ecological conservation and environmental protection serve as an important driver to advance this magnificent journey in the past 40 years.

But how to define ecological conservation and environmental protection? Before 1978, China was still a country featuring low industrialization and slow urbanization. At that time, China gave top priority to food supply and crop production, but at the cost of ecological degradation. Furthermore, like Western nations, China then was also faced with the environmental challenges brought by its own industrialization. Understandably, the shocking lessons from the West and its own practices made China highly vigilant of the possible environmental threats, even before the reform started. The ecological balance quickly grew into a major concern of Chinese society during the early stage of reform and opening-up. Later, as the industry expanded and cities sprawled, ecological damage was gradually reduced but pollution was on the rise. In the twenty-first century, Chinese economy, which has been in the course of globalization, shows typical characteristics of heavy chemical industry: high-energy consumption, high-energy demand and high pollution. Measures for ecological balance and pollution control are no longer enough to answer the needs of sustainable development. Thus, against this backdrop, China launched an initiative to build an ecological civilization that consists of high productivity, prosperous life and ecological soundness. First, technologies are encouraged that can save resources, control pollution and conserve ecosystem are encouraged. Second, moves at institutional level are taken, including the promulgation of new laws and regulations, setting higher standards, and assessing performances. Third, market measures are adopted, ranging from tax incentives, price guidance, to transfer payment. China started a new age where a five-sphere integrated plan, which includes the development of economic, political, cultural, social and ecological undertakings, was introduced, with ecological civilization being put in the most prominent position and integrated into every aspect of economic and social development.

Though having achieved environmental feats, the past 40 years of reform and opening-up have also been a period of time when China rose to huge environmental pressures and challenges with effective countermeasures. But how can we summarize the experience and stories therein? We have selected, first, the regulatory areas like laws, regulations, policies and planning; second, the core fields such as ecological protection, circular economy and pollution control; and third, areas that are restrictive and policy-oriented, for instance, water resources and renewable resources, as well as the major areas that are in the process of transformation, including climate change, sustainable development and the building of ecological civilization. Written by leading experts who have long engaged in China's ecological conservation and environmental protection, this book provides a holistic and systematic review and summary of the related areas in terms of their historical process, characteristics at different development stages, intrinsic motivation, actions taken, performances achieved, trends, future trends and so forth. On top of that, the authors present us with much insightful thinking, academic analysis, research on rules, case studies and future visions.

The journey and stage attributes of ecological conservation and environmental protection in China coincide with the timing of the major events of the country's reform and opening-up, with each juncture signifying an iconic event as well as the beginning of a new period. The first juncture is undoubtedly in 1978, the first year China adopted the reform and opening-up policy. The second juncture is in 1992, when Deng Xiaoping delivered his Southern Tour Speeches. The third juncture is at the end of 2001, the end of which witnessed China's entry into the WTO and the beginning of its all-round, large-scale and in-depth globalization. The fourth juncture is in 2012, the year China entered a "new normal" period featuring the transformation of development patterns. Surely, ecological conservation and environmental protection in China have its own characteristics, and its advancement has innate rules too. For example, it has been highly associated with major global agendas, like the UN Conference on the Human Environment (1972), the UN Conference on Environment and Development (1992) and the UN Conference on Sustainable Development (2012), all of which matched the course of China's reform and opening-up. The year 2015 saw the adoption of the 2030 Agenda for Sustainable Development and the Paris Agreement by the United Nations, which were supposed to be reached in 2012 but were not signed until three years later due to the severe differences between global south and global north and the major changes in the world's economic and social development. Some big environmental events in China, though not necessarily chronologically identical to the above-mentioned time, also marked the country's environmental feats achieved during four decades of reform and opening-up. Therefore, the book does not divide the timeline of China's environmental advancement in a stringently unified way.

When writing and revising this book, all the contributing authors held themselves, with rigor, accountable for the time and quality, making the publication fully conform to the deadline requirement of the publisher. Mr. Xie Shouguang, president of Social Sciences Academic Press (China), participated in the planning of the book, offering lofty ideas and visions. Editors Li Yanling and Shi Xiaolin exchanged views conscientiously with contributing experts in their editing work. He Ni, a researcher assistant with CASS Think Tank for Eco-civilization Studies, provided quality service too by taking the initiative to contact all the authors and answering their phone calls around the clock. While we have tried to make the book as intellectually rigorous as possible, there are surely many ways in which it can be made better. We welcome suggestions from our readers.

Pan Jiahua Zhuang Guiyang
October 2018

Contents

List of Figures xi
List of Tables xiii

Chapter One: "Protecting the Environment Pays": Theory and Practice 1
    Pan Jiahua and Zhuang Guiyang

Chapter Two: Environmental Policies: Evolution and Evaluation 19
    Zhou Hongchun

Chapter Three: Environmental Protection Planning: Review and Outlook 51
    Wu Shunze, Wan Jun, Yang Liyan and Zhao Zijun

Chapter Four: Circular Economy 89
    Qi Jianguo, Wang Yingjie and Ma Xiaoqin

Chapter Five: Water Resources: Exploitation and Protection 119
    Xia Jun and Zuo Qiting

Chapter Six: Adapting to Climate Change: Policies and Actions 155
    Zhuang Guiyang and Bo Fan

Chapter Seven: Sustainable Development: Strategy and Practice 187
    Chen Ying

Chapter Eight: Sustainable Urban Development: Practice and Experience 211
   Wang Mou, Kang Wenmei, Liu Junyan, Lv Xianhong, Zhang Ying and
   Luo Dongshen

Chapter Nine: Toward an Ecological Civilization 247
   Li Meng

List of Figures

Figure 3-1	China's Economic Growth Rate since the Reform and Opening-up	53
Figure 3-2	Roadmap of China's Environmental Planning and Socioeconomic Development Stages	58
Figure 5-1	Water Resources Utilization and Protection in China from 1978 to 2018 and Prospect of Future Development	140
Figure 6-1	Decoupling between China's economic growth and carbon emission	161
Figure 7-1	Changes in China and the world's average human development index from 1990 to 2013	198
Figure 8-1	Overall process of sustainable urban development in China	217

List of Tables

Table 3-1	The Positioning by the National Environmental Protection Conferences (NEPC) of the Role of Environmental Protection in Economic and Social Development	55
Table 3-2	List of Key Eco-environmental Protection Plans since the Reform and Opening-up	76
Table 5-1	Development Stages and Representative Events of Water Resources Utilization and Protection in China	123
Table 6-1	Energy conservation and emission reduction constraint goals in five-year plans and their fulfillments	162
Table 7-1	China's implementation of the MDGs	197
Table 8-1	Proposal time and practice of different types of pilot and demonstration cities	233

CHAPTER ONE

"Protecting the Environment Pays": Theory and Practice

PAN JIAHUA AND ZHUANG GUIYANG[*]

1. INTRODUCTION

Chinese economy has been greatly rejuvenated, thanks to the reform and opening-up since 1978, dwarfing the whole world in growth rate with an ever-rising size. However, the four decades of economic take-off is also a period of time when the country wound out of all the twists and turns in environmental protection, and a space of time when environmental deterioration in general was improved by a large margin. The past forty years of rapid industrialization and urbanization in China have not been able to escape from the outburst of environmental woes that Western countries experienced during their century-long industrialization. The country, consequently, has entered a critical period with overlapping pressures and heavy burdens. After forty years of unremitting efforts, China has made huge progress in the institutional establishment for environmental protection, improving ecological conservation to a new stage of development. Generally speaking, China's environmental movement did not advance in an isolated way, for it has always been closely associated with the country's economic

---

[*] Pan Jiahua: Ph.D. in Economics, University of Cambridge. Member of CASS and Director, Research Fellow and Doctoral Supervisor of the Institute for Urban and Environmental Studies, CASS; Zhuang Guiyang: Ph.D. in Economics, Research Fellow and Doctoral Supervisor of the Institute for Urban and Environmental Studies, CASS.

development, reform and opening-up. In the early 1980s, China's policy was to "sell mineral resources as fast as possible" and to "live on what the land and sea can offer." While in the rapid industrialization and urbanization, China began to realize that it needed not only golden mountains but also clear water and lush mountains. Since 2010, under the guidance of this environmental philosophy, Chinese people have begun to refused to trade their lucid waters and lush mountains just for economic benefits. Recent years have seen the emergence of a new age of ecological civilization featuring the harmonious symbiosis between nature and humans. As we can see, along with the special junctures in the forty years of socioeconomic development, China's environmental protection also displayed different characteristics in its development, policies and institutional establishment.

2. LOCAL POLLUTION AND THE BEGINNING OF INSTITUTIONAL IMPROVEMENT (1978–1991)

The Third Plenary Session of the Eleventh Central Committee of the Communist Party of China, held in 1978, marked the beginning of China's reform and opening-up in an all-round way and its exploration of socialism with Chinese characteristics through learning while moving forward, or figuratively speaking, "crossing a river by feeling the stones." The Third Plenary Session made clear that the work of the whole Party and the whole country should focus on socialist modernization drive. With this shift in the national development strategy, China started to establish its institution for environmental protection by drafting relevant laws, regulations and policies.

In 1972, Chinese delegation participated in the UN Conference on the Human Environment. In 1973, the first National Environmental Protection Conference approved the environmental protection policy of "overall planning and rational distribution; utilizing resources comprehensively and turning hazards into benefits; relying on the people and involving everybody; and protecting environments for the benefit of the people," which is also known as the "Thirty-Two-Character Principles." In a very short time, the Leading Team for Environmental Protection, along with its office, was officially established under the State Council. Governments at lower levels were urged to establish their environmental agencies correspondingly. All these efforts officially symbolized the outset of Chinese environmental movement in modern times centering on pollution prevention and control. Reform and opening-up needed legal safeguard, but at that time, the legislative process had to start all over with so many things to attend to. Still, the legislation for environmental protection was written into the agenda. In 1979, the *Environmental Protection Law of the People's Republic of China* was officially promulgated, marking the official inclusion of environmental protection into the legislation. In 1983, the second National Environmental Protection Conference made environmental protection a basic state policy and established eight "synchronous development" guidelines.[1] In 1989, at the third National Environmental Protection Conference, Chinese government further proposed the "three major policies" and "eight major management systems,"[2] which produced far-reaching influence on China's environment movement.

Environmental protection was written into the Sixth Five-Year Plan (1981–1985) as an independent chapter. Since 1983, environmental protection has been an integral part of the annual report on the work of the government, serving as a guarantee for the implementation of environmental projects and objectives. Environmental protection agency in China has gone through different stages of development too. In 1982, the Office of the Leading Team for Environmental Protection under the State Council, a previous temporary institution that had been established for ten years, was transformed into a regular permanent institution, while at the same time, the Environmental Protection Bureau was set up under the Ministry of Urban and Rural Development and Environmental Protection. In 1984, due to the fact that the environmental protection involved the coordination of different departments, the State Council established the Environmental Protection Commission and changed the name of Environmental Protection Bureau to National Environmental Protection Agency (NEPA, a sub-ministerial level agency), which was still under the leadership of the Ministry of Urban and Rural Development and Environmental Protection. In 1988, the State Council reorganized its institutions and severed the NEPA from the Ministry of Urban and Rural Development and Environmental Protection. The NEPA was hence defined as the competent department directly under the State Council for integrated environmental planning, monitoring and management.

By the end of 1991, China had drafted and promulgated 12 laws on resource and environment, more than 20 administrative laws and regulations, over 20 departmental regulations, 127 local laws, 733 local regulations and a large number of normative documents. Thus, the legal system of environmental protection had been preliminarily formed, laying a legal foundation for strengthening environmental administration.

In general, environmental protection developed in parallel with economy, and to certain extent, even beyond the corresponding phases of economic growth in terms of the building of management system. At that time, China, a country still practicing central planning to a large extent, was highly vigilant of repeating the development mode of "pollution first and management later" of western countries, but was equally desirous of the development road of "protection first and development later." Chinese government became more alert in the 1980s when economic growth and industrialization, especially the rise of enterprises in small towns, led to ecological damage and increasing pollution.[3] However, before 1992, Chinese economic development was basically in the exploring period of "crossing the river by feeling the stones." Most pollution accidents emerged during this period of time, mostly were local or of point-source pollutions.

3. WORSENING ENVIRONMENTAL POLLUTION AND FASTER LEGISLATION (1992–2001)

Since China began to accelerate and expand its reform and opening-up in 1992, the year that marking a turning point in Chinese history, the country experienced two decades of rapid economic development and the highest environmental stresses ever as well. During his tours in Shenzhen and Zhuhai, Deng Xiaoping delivered two very important speeches, which played a critical part in propelling China's economic reform and social development in the 1990s. People were able to take their first bite into the benefits of thriving economy and saw the hope of being lifted out of poverty. However, ecological and environmental integrity was sacrificed for economic growth in many places as China then was still in the transition from planned economy to market economy, a period dominated by persistent extensive growth mode, which had resulted in serious pollution and ecological loss as well as threats and damage to people's livelihood and the country's sustainable social and economic development.

After Deng Xiaoping's visits to southern China in 1992, the country ushered in a new round of economic boom. But as a Chinese proverb goes, "turnips are not washed when selling fast," which means, in this case, that environmental health was compromised for economic gains. This situation was complicated by the robustious and disorderly development of township enterprises in the 1980s, giving rise to graver environmental pollution. For example, some rivers and lakes turned black and malodorous, which, along with blue algae eruption, jeopardized the drinking water safety. Smog enshrouded many cities, directly causing the outburst of respiratory diseases among urban residents. To make it worse, pollution, which began to expand from cities to the countryside, from eastern China to western regions and from local to countrywide in scale, not only led to the increasing number of environmental accidents but also grew into the hotspots of social complaints. Worse still, pollution grew as air and soil pollution were added to water pollution, causing widespread concern. Urban environmental pollution was especially alarming, which was largely caused by the high population density and over-concentrated manufacturing enterprises, unsound industrial structure and building layout, underdeveloped infrastructure for urban environmental protection, and the concentrated discharge of residential waste and industrial wastes (gases, water and solid).

But at that time, it was only possible to tackle the pressing environmental issues, among which the case of Huaihe River was the most typical. In 1989 and 1994, an accident occurred and resulted in the serious pollution of Huaihe River, challenging more than 1.5 million residents in Anhui Province and Jiangsu Province in their access to drinking water. As a response, the State Council laid down the objective that Huaihe River should be cleared before the end of the twentieth century, symbolizing the beginning of water treatment for Huaihe River. In Huaihe River basin alone, Chinese government closed down 999 small paper mills and renovated 1,139 polluting enterprises from 1994 to 1998, achieving initial results. The achievement, however, was not consolidated, despite the fact that the water quality of some major monitoring sections of Huaihe River was improved obviously, with some reaching Grade III. Due to the inadequate industrial restructuring, some small yet still polluting enterprises started their business again, leading to the rebound of the Huaihe River pollution. This is largely because the environmental impairment outgrew pollution abatement during that period of time, which was more known as the time of environmental debt accordingly.[4]

To beef up environmental administration, the *Decision of the State Council on Several Issues Concerning Environmental Protection*, which was promulgated in 1996, stipulated that fifteen kinds of heavily polluting small enterprises to be closed, in an attempt to contain serious pollution sources infamous for backward technology, little hope of abatement, large number and huge amount of discharge. The shutdown of the fifteen kinds of polluting small enterprises[5] put an end to the irrational development of township enterprises since the 1980s, which could be described as "every village developed heavy industry and every household had a smoking factory." Thanks to the restructuring and upgrading, effective control was maintained over the pollution of atmosphere, water and the heavy metal in soil, which, to a certain extent, protected the health of workers who used to work in these township enterprises.

During this period, China strengthened step by step, its legislation for environmental protection. The Environment and Resources Protection Committee, which was created by the National People's Congress as the Environmental Protection Committee in 1993 and changed into the current name in 1994, formulated five laws, including the *Law on Promoting Clean Production* and the *Law on Environment Impact Assessment*, while revising three laws such as the *Law on the Prevention and Control of Atmospheric Pollution* and the *Law on Prevention and Control of Water Pollution*. Thus, China established a legal system for environmental protection which consists of eight environment laws, fifteen natural resources laws, more than fifty administrative regulations, nearly 200 departmental regulations and normative documents and over 1,600 local environmental laws and regulations.

The UN Conference on Environment and Development held on June 12, 1992 prompted China's *Ten Countermeasures for Environment and Development*. The year 1994 saw the promulgation of *China's Agenda 21*, which is also known as the *White Paper on China's Population, Environment and Development in the 21st Century*, an overall strategy, plan and countermeasure scheme for sustainable development drafted by China in accordance with the Agenda 21 adopted by the United Nations at Rio in 1992. It is a guiding document for the Chinese government to formulate medium- and long-term plans for national, economic and social development. In 1995, besides giving the environmental protection a status as a basic state policy, China further established sustainable development as a national strategy with the same importance as that of invigorating China through science and education.

In 1998, the State Council reconfigured its institutional structure, renaming the Environmental Protection Bureau as the State Administration of Environmental Protection, elevating the agency's rank in the government, and suspended the Environmental Protection Commission. The China Council for International Cooperation on Environment and Development (CCICED) was founded in 1992 to absorb international suggestions on economic and environmental affairs and reach out to world community on China's environment policies and attitudes.

Generally speaking, the first twenty years since 1992 was a course of rapid growth—a double-digit annual growth rate basically, which, when interwound with industrialization and urbanization, exerted ever graver pressure on the carrying capacity of ecosystem, whose upper limit was approached step by step to the maximum. It is largely the result of the poor environmental administration at that time. Due to the fact that in 1993 China established the socialist market economy as its basic economic system, the focus then was shifted to economic growth, which deprived the environmental protection system of its rigid constraints on one hand and made the environmental enforcement more flexible on the other hand. As the pressure on economic development increased, the requirements on environment protection were loosened, thus leading to the imbalance between the two. Objectively speaking, environmental protection, whose intensity was far from enough because of the insufficient governance capacity and excessively powerful market forces, became increasingly weak if compared with the environmental rules necessary for rapid economic growth.

China's environmental protection was in a very grave situation at that time as the accelerating industrialization and urbanization then highlighted the striking tension between economic growth and environmental protection. The severe environmental pollution and ecological deterioration in some areas could be concluded from the excessive discharge of major pollutants beyond the environmental carrying capacity, the serious pollution of water, air and soil, an increasing amount of solid wastes, automobile exhaust, persistent organic pollutants and so forth. The environmental problems that had appeared by stages in the century-long industrialization in western countries came to China all at the same time, giving rise to the increasingly prominent conflicts between environment and development. The major problems in China's development included the relative shortage of resources, fragile ecosystem and inadequate environmental capacity.

4. CONTAINING ENVIRONMENTAL DETERIORATION AND STRENGTHENING GOVERNANCE (2002–2011)

China became the 143rd member of the World Trade Organization (WTO) on December 11, 2001, following a fifteen-year negotiation process. It signified China's holistic integration into the world economy, the beginning of another round of economic boom and of the search for a new road to industrialization. From 2001 to 2010, China's GDP grew at an average annual rate of 10.5%, faster than the 9.3% in the 1980s and 10.4% in the 1990s. China became the second largest economy in the world after its economic aggregate overtook Japan in 2010.

China proactively followed the accelerating trend of international division of labor. By giving full play to its comparative advantages, China undertook international industrial transfers and launched an export-driven outward-oriented economy. Great efforts were spent on the development of foreign trade and the promotion of two-way investment, as a result of which China's open economy began to run on a fast lane with overall national strength being enhanced on a constant basis. Thanks to the stimulation from the upgrading of consumer consumption structure, the capital- and technology-intensive industries, such as the automobile industry, achieved rapid development and became the country's main industries, which propelled the expansion of related sectors like steel and machinery.

However, during this period, China failed to fundamentally check the trend of overall environmental deterioration, with conflicts in environmental protection being more visible and pressure constantly on the increase. For example, pollution became increasingly protruding in some key river basins, sea areas and rural areas. Just as striking was the pollution caused by heavy metals, chemicals and persistent organic pollutants as well as that of soil and groundwater, not to mention the serious smog in some regions and cities. In some areas, the discharge of major pollutants exceeded the environmental capacity while some other places saw terrible ecological damage, degraded eco-functions and fragile environment. Nuclear and radiological risks also increased. The high incidence of environmental emergencies made environmental problems a threat to human health, public safety and social stability. At the same time, environmental constraints on economic growth grew progressively powerful along with the continuous expansion of population, the rapid progress of industrialization and urbanization, the constant rising energy consumption and pollutant production. Global environmental issues such as climate change and biodiversity conservation were also exerting ever bigger pressure.

Qu Geping, former minister of National Environmental Protection Agency and president of China Environmental Protection Foundation, said, "Truly, it took China just thirty years to finish the industrialization which took the West more than two hundred years. However, it also took China only thirty years to be as polluted as the West had been for more than one hundred years." "The development of heavy chemical industry since 1992 signified that China entered a period with the highest environmental pressure."[6] Especially between 2002 and 2012, China built a large number of heavy chemical projects, most of which were high in pollution, emissions and energy-intensive, including steel, cement, chemicals and coal power. Coming together with the large-scale development of the heavy chemical industry was the huge demand for and pressure on energy and resources as well as serious pollution. Although China began to implement energy conservation and emission reduction since 2006, heavy chemical industry continued to expand, with the discharge of contaminants still on the rise and difficult to curb. Then, the government had to step up its efforts on emission reduction and introduce market-based measures. However, although major pollutants did decline gradually, the environmental quality was just as bad, with a high incidence of pollution accidents, which further triggered frequent public opposition.

At the same time, China was under huge pressure from international community in terms of climate change. Since its entry into WTO in 2001, China began to integrate into the world economically in an all-round way. The increasingly grave environmental pollutions, along with the hiking energy consumption and carbon emission, aroused wide international concerns. China was gradually looked upon as a threat to environment. Following closely on the heels of the industrialization of the twenty-first century's powerhouse and its development of heavy industry was a sharp rise in its total amount of CO_2 emissions. With a notable catch-up effect, it took only three years for China to overtake the USA and the EU in carbon emissions. In 2003, China's CO_2 emissions reached 4.052 billion tons, exceeding for the first time the aggregate emissions from twenty-eight EU countries, which stood at 3.942 billion tons. In 2006, China became the largest carbon emitter with an emission of 5.912 billion tons, compared to 5.602 billion tons by the USA. Then in 2012, China surpassed the USA and the EU combined in carbon emissions. It was not a surprise, however, as China then, with the world's largest population, had grown into a world factory which used coal as its major energy source. Therefore, when measured on a country basis, China should be indeed held accountable.[7]

In the face of growing carbon emissions, China became more aware of its responsibilities for emission reduction and introduced a series of measures accordingly. For example, the Eleventh *Five-Year Plan* for Social and Economic Development adopted as a binding indicator the reduction of carbon emission per unit of GDP, which was further written into the Twelfth *Five-Year Plan* as a major indicator for low-carbon development. All these moves fully demonstrated the political will and determination of the Chinese government in energy conservation and emission reduction.

The beginning of the twenty-first century saw the adoption by the CPC Central Committee and the State Council of the Scientific Outlook on Development, which stipulated that all development be people-oriented, comprehensive, coordinated and sustainable. At the National Symposium on Population, Resources and Environment held in 2003, Hu Jintao, General Secretary of the CPC Central Committee, stressed that environment protection should ensure public access to clean water, air and safe food so that people could work and live in a good environment, and that top priority be given to solve the striking issues that compromise public health. The national symposiums on population, resources and environment in the next two years (2004–2005) both set the people's livelihood as a target of environmental protection. In 2008, the State Council reorganized its institutions and upgraded the National Environmental Protection Agency to the Ministry of Environmental Protection, an institution of ministerial level. Its main responsibilities were to "formulate environmental protection plans, policies and standards, organize the implementation of these plans, policies and standards, organize and prepare environmental functional district divisions, supervise and manage the prevention and control of environmental pollution, and coordinate the resolution of major environmental problems."

During the Tenth *Five-Year Plan* period, the Chinese continued to intensify their environmental efforts. They eliminated a number of backward production capacities notorious for high consumption and high pollution, accelerated pollution control and the construction of environmental infrastructure in urban areas with greater efforts on the environmental management in key areas, river basins and cities. However, the targets of the environmental protection plan of the Tenth *Five-Year Plan* were not met in full. Compared with the year 2000, the emission of sulfur dioxide increased by 27.8% in general and by 2.9% in SO_2 Pollution Control Zone and Acid Rain Control Zone, with the emission of soot up by 1.9% and that of industrial dust down by 16.6%, and the chemical oxygen demand decreased by only 2.1%. The failure to fulfill the indicators of the Tenth *Five-Year Plan* gave rise, to a large extent, to the environmental restrictive indicators. The Eleventh *Five-Year Plan* set the goal of reducing energy consumption per unit GDP and the total emission of major pollutants by 20% and 10%, respectively, while increasing the forest coverage rate to 20%. During the

Eleventh *Five-Year Plan* period, important progress was made in policies, public awareness, institutional establishment and capacity building, as a result of which, all the environmental targets and key tasks were fully fulfilled. It was especially true when it came to energy conservation and pollution reduction, the two targets that exceeded the planned requirements. On the basis of previous achievement, the Twelfth *Five-Year Plan* laid down new goals for energy conservation, carbon reduction and the reduction of the total amount of pollutants.

In general, compared with the previous decade between 1992 and 2001, the trend of environmental deterioration had not been fundamentally curbed from 2002 to 2011. Environmental contradictions became more noticeable, threatening the carrying capacity of ecosystem to its upper limit. In 1992, China decided to build a socialist market economy system through reforms. In 2001, China joined the WTO, which was deemed as a milestone effort toward further opening-up. Admittedly, reform and opening-up brought great impetus to economic growth. It was, however, the same economic growth, or in other words, the growth rate of GDP, investment scale, and fiscal and tax revenue, that long constituted the core of performance assessment indicators in many places, for they could reflect the economic quantity and growth rate. The GDP also had the most powerful say when it came to the evaluation and appointment of local officials, as a higher GDP growth rate meant a shortcut to promotion. As a result, in order to pursue high growth rate, many officials built large showcase projects which could prop up their own political image at the cost of environment. Many of these projects were developed against the rules of rational development. This type of GDP-oriented assessment system greatly undermined the public will to combat the wrong development mode of "pollution first and treatment later." Industrialization and urbanization brought ever-growing environmental pressure that constantly tested the bearing limit of the ecological system. Despite that China made great headway in improving the eco-environment protection system in this period, including the upgrading of the State Environmental Protection Administration to Ministry of Environmental Protection, the rapid eco-environmental deterioration was not stopped and eco-environmental pollution events were frequently recorded.

5. POLLUTION BATTLED COMPREHENSIVELY AND ENVIRONMENTAL QUALITY STEADILY IMPROVING (2012–PRESENT)

The four decades of reform and opening-up have witnessed China's fast socioeconomic progress and the significant improvement in people's lives. The country has built a moderately prosperous society in an all-round way and entered a new era of socialism with Chinese characteristics. Currently, the principal imbalance facing Chinese society is that between the unbalanced and inadequate development and the people's ever-growing needs for a better life. Such needs definitely include the one for a good environment, which, however, can barely be met, for the present environmental protection and ecological restoration still lag behind the economic and social development in China, which is a prominent respect in the "unbalanced and inadequate development."

The eighteenth CPC National Congress has made clear that it is vital to vigorously build an ecological civilization and a beautiful China and realize the sustainable development of the Chinese nation. Over the past five years, the concept that "lucid waters and lush mountains are invaluable assets" has been deeply rooted in the minds of people. The top-level design and institutional construction of ecological civilization have been accelerated, which puts pollution control on a fast track. The green development has achieved remarkable results, with the ecological quality being improved continuously. The picture of a new beautiful China is unfolding.

Since the Eighteenth CPC National Congress held in 2012, the CPC Central Committee, with comrade Xi Jinping at its core, has taken ecological conservation as an important content for the promotion of economic, political, cultural, social and ecological progress and of the "four comprehensives" strategy, i.e., to comprehensively build a moderately prosperous society, deepen reform, implement the rule of law and strengthen Party self-discipline. Ecological conservation is seeing historic and holistic changes and turning points, thanks to the fundamental, pioneering and forward-looking work planned by the CPC central committee. Also, since the eighteenth CPC National Congress, remarkable results have been achieved in ecological conservation. The consciousness and initiative of implementing the concept of green development have been significantly enhanced throughout the country. Environmental protection should no longer be neglected. This is not only a positive response to the demands of the public but also a concrete embodiment of the responsibility taken by the CPC Central Committee with Xi Jinping at its core for the sustainable development of the country and the people.

Another noticeable improvement in the ecological conservation was the so-called "War to Defend the Blue Sky." Compared with 2013, the average concentration of PM10 in 338 cities of prefecture level and above dropped by 22.7% in 2017.[8] All the targets stipulated in the "Action Plan on Pollution Prevention and Control" in China had been reached. The proportion of good quality water over surface water monitoring points kept increasing, while the percentage of Class V water bodies dropped to 8.3%. The black and odorous water bodies in thirty-six key urban built-up areas were basically eliminated. A comprehensive soil contamination survey was carried out along with the demarcation of basic farmland. The biosafety disposal rate of domestic wastes reached 97.14% in urban areas. 74% of administrative villages had access to residential waste disposal.

The ecological protection and restoration have progressed smoothly. China is now steadily implementing a number of major ecological protection and restoration projects, including the protection of natural forest resources, returning farmland to forest and grassland, returning grazing land to grassland, the construction of shelterbelt system, the protection and restoration of rivers, lakes and wetlands, sand control, soil and water conservation and the treatment of stony desertification, the protection of fauna and flora, and the construction of natural reserves. Land desertification and stony desertification both declined in areas for the past three consecutive monitoring cycles. The grassland, which covers an area of nearly 400 million hectares, or 41.7% of the land area, is now the largest terrestrial ecosystem in China. By the end of 2017, the total area of nature reserves of various types and levels reached 1,471,700 square kilometers.

The intensity of energy and resource consumption was greatly reduced, thanks to the industrial restructuring and the removal of excess capacity. De-capacity efforts yielded initial results in key industries. From 2013 to 2017, the steel industry saw the reduction of steel and coal production capacity by 170 million tons and 800 million tons, respectively. Also weeded out were small coal-fired boilers in urban built-up areas at or above the prefecture level as a result of tighter control over bulk coal, with 71% coal-fired power units achieving ultra-low emission. On top of the improvement of fuel quality and the termination of heavily polluting vehicles, the government promoted the sales of 1.8 million new-energy vehicles. In 2017, the proportion of clean energy consumed increased to 20.8%, with energy and water consumption per unit GDP both dropping by more than 20%. China is No.1 in the world in terms of the installed capacity of hydropower and wind power, the size of nuclear power under construction and solar heat collector area.

The legal and institutional framework for ecological civilization has been built gradually. The CPC Central Committee and the State Council issued the *Opinions on Further Promoting the Development of Ecological Civilization* and the *Overall Plan for the Reform of Ecological Civilization System*. The National Development and Reform Commission released the *Green Development Indicator System* and the *Assessment Target System for the Development of Ecological Civilization*. All these documents served as the basic framework for the development of ecological civilization. Related laws and regulations, including the *Environmental Protection Law*, the *Law on the Prevention and Control of Atmospheric Pollution* and the *Ambient Air Quality Standards* had been revised. The system of setting chiefs for rivers, lakes and bays was also introduced. The environment assessment and examination systems were also improved on a constant basis. Pilot projects for national parks, low-carbon city and sponge city were carried out in full swing, complementing the top-level design of ecological conservation.

As the intensity of carbon emissions continues to decline, China leads the international cooperation on climate change. The carbon intensity reduction rate has been included into the evaluation statistics of national economic and social development in Thirteenth *Five-Year Plan* period, which to some extent, inevitably restrained economic growth. Compared with 2005, China's carbon intensity fell by 42% in 2016, fulfilling the target of a reduction to 40–45% by 2020, which was promised at the 2009 United Nations Climate Change Conference in Copenhagen.[9] China proactively pursued the national strategy for addressing climate change and issued the *National Plan on Climate Change (2014–2020)* and the *National Plan on Implementation of the 2030 Agenda for Sustainable Development*. After the US withdrew from the *Paris Agreement*, China firmly stated that it would continue to fulfilling its commitment to emission reduction and pursuing the goals in Paris Agreement with determination and tangible results.

After 2012, the CPC Central Committee launched a new round of reform and opening-up, putting the development of ecological civilization on top agenda. General Secretary Xi Jinping pointed out that China's environmental quality continued improving and demonstrated a steady and sound trend, but the results were not solid enough. Currently, the development of ecological conservation is not only in a critical period filled with multiple pressures, a moment as vital of providing more quality ecological goods to meet people's ever-increasing needs for a beautiful environment but also a window period capable of solving prominent ecological problems. The economic new normal (which means the slowdown in growth rate) has created, to some extent, an important "window period" for environmental protection. Also, market saturation (excess capacity) means we could set more stringent standards on environmental protection. Now, we have the capacity and conditions to solve ecological and environmental problems, thanks to the growth of financial resources, technological progress and the accumulation of experience.

5. HARMONIOUS COEXISTENCE BETWEEN HUMAN AND NATURE AND PURSUING ECOLOGICAL CONSERVATION IN AN ALL-ROUND WAY

Since the eighteenth National Congress of the Communist Party of China, General Secretary Xi Jinping has reiterated that "lucid waters and lush mountains are invaluable assets," a philosophy so important that it was written into the report of the nineteenth National Congress of the Communist Party of China, along with the content that ecological conservation is a plan of long-term significance for the sustainable development of the Chinese nation. The concept that "lucid waters and lush mountains are invaluable assets" was further incorporated into the Constitution of the Communist Party of China, which serves as a powerful political guarantee for the achievement of ecological conservation in the new era. Thus, a new modernization drive featuring the harmonious coexistence and development of man and nature is taking shape.

In order to better develop ecological civilization in the new era, for one thing, "six principles" are adopted, including the basic principle of harmonious coexistence between humanity and nature, the development philosophy that lucid waters and lush mountains are invaluable assets, the tenet that a good environment is in the common interest of people, the systematic thinking that mountains, forests, rivers, lakes and grasslands are the shared living community, the firm determination to protect environment with the strictest system and according to the rule of law, and the accountability that a great country should hold for building an ecological civilization together with global community. For another, solving the most prominent ecological problems, as a matter concerning people's life and work, is given top priority, for the prevention and control of pollution, is a battle too tough, important and difficult to lose. First, we put the improvement of air quality as the most fundamental requirement, with the heavily polluted weather basically eliminated and the blue sky successfully defended. Second, we guaranteed the safety of drinking water by implementing the action plan for the prevention and control of water pollution and clearing the black and odorous water bodies in urban areas. Third, we strengthened the management, control and restoration of polluted soil to ensure that people could have safe food and living environment. Fourth, we endeavored to build beautiful countryside by improving the human settlements in rural areas on a constant basis.

In the course of building ecological conservation for a beautiful China, the fundamental solution is to promote green development in an all-round way. The report of the Nineteenth National Congress of the Communist Party of China also attaches great importance to the harmony between human and nature, and lays down major tasks of building a beautiful China by "promoting green development." In essence, green development, as one of the five development concepts, is to solve the problem of harmonious coexistence between man and nature and highlights that man and nature form a community of life. But the green development in building a community with a shared future for mankind requires the developed countries to take the lead, on one hand, and the developing countries, on the other hand, to be innovative and adopt a development model featuring low consumption and low emission instead of the old road of high pollution and high emission of the western countries.

Being neither simple nor superficial, green development means to promote ecological conservation in an all-round way, which requires us to respect nature, follow its ways and protect it. Green development calls for technological innovation and revolution, through which we can save water by recycling, conserve energy by improving efficiency, spare material by reducing consumption and manage land by using it intensively. Revolutionary technologies can lead to changes in quality or by magnitude. Likewise, green development needs a revolution in consumption, which features a simple, moderate, green and low-carbon lifestyle instead of an extravagant, excessive and irrational consumption model. In addition, green development in production and consumption means to develop an outlook on ecological conservation, the awareness of which would not be cultivated without people's initiative as well as the legal and policy regulation and guidance. The Nineteenth National Congress of the Communist Party of China proposes to enforce stricter pollutants discharge standards and see to it that polluters are held accountable. The implementation of these requirements is conducive to the green development in all respects.

In an article published in Zhejiang in *A New Vision for Development* of *Zhejiang Daily* on March 23, 2006, Xi Jinping said that people's understanding of the concept that lucid waters and lush mountains are invaluable assets went through three stages. In the first stage, people traded environment for wealth at any cost, with no or little consideration given to the carrying capacity of environment. In the second stage, people wanted both environment and wealth, but the clashes between economic development and the lack of resources and environmental deterioration began to emerge. People came to realize that environment is the foundation of our survival and development and only by keeping a good environment can we have sustainable development. In the third stage, people began to appreciate that lucid waters and lush mountains are invaluable assets per se and can bring wealth continuously. The evergreen environment is a ready source of wealth, for ecological advantages can be turned into economic benefits when they coexist in unity and harmony.

Xi Jinping's incisive exposition, which gives top priority to ecological conservation, embodies the unity of economic development and environmental protection, where lies the inherent logic that ecological advantages are economic benefits. The concept that lucid waters and lush mountains are invaluable assets conforms to natural, social and economic rules, thus is of great theoretical value and practical significance. Green development and economic growth are by no means opposites. On the contrary, reform and innovation can efficiently motivate such factors as land, labor force, capital and natural scenery, among many others, and turn natural resources into ecological assets and economic benefits.

Ecological assets refer to natural resources that can provide humans with tangible and intangible benefits, including water, air, land, fossil fuel and various ecosystems formed by basic ecological elements. The value of ecological assets is expressed in the form of the value of ecological service, of resources and energy, and of ecological goods produced by human activities. Ecological assets first become ecological capital through the development and investment from human society, which liquidize assets, and second, ecological goods through operation and finally have their value realized through the market. The constant changes of the form and value of ecological assets lead to the continuous appreciation of ecological assets.[10]

The key mission and symbol of the transformation and amelioration of industrial civilization is to establish a system of ecological conservation, in the way as follows. First of all, faster pace is needed to build and refine the eco-culture system based on ecological values. Only by respecting and following the rules of nature can we humans strike the balance between using nature sustainably and coexisting with nature. Second, a system for ecological economy, which mainly features the ecologicalization of industries and the industrialization of ecology, is to be built. In any society and at any time, the goal of social development would not be realizable, nor would it be possible to effectively protect the environment without economic vitality and the support of material wealth. However, environment protection, which is economic development in essence, cannot sustain if it means only input but produces no output. Solar energy, for example, is a substitute for fossil energy, but it is also an industry that provides jobs, generates economic returns and offers energy services. Urban mines, as it is called, are to turn the polluting wastes into resources and to industrialize the ecology. Third, it is important to establish a results-based accountability system with the improvement of environmental quality as the core. The quality of environment will only be guaranteed when the targets are clearly defined and assigned to each stakeholder and supervision is in place. Our society is desperate for an environment of high quality, which means blue sky, lucid water and clean soil. Green and beautiful environment and rich biodiversity will not be possible without the synergy between mountains, rivers, lakes, grasslands and fields. Fourth, it is necessary to construct a system for ecological civilization which takes the modernization of governance system and capacity as the safeguard. Ecological preservation according to the rule of law and social participation are the basic connotation of the modern governance system of ecological conservation. Fifth, it is imperative to build an ecological security system with the benign cycle of ecosystem and the effective prevention and control of environmental risks as the focus. Natural system changes all the time, partly due to human interventions, but the key is to efficiently control and manage all the natural and artificial risks and events that might jeopardize environment. Only in this way can the system for ecological economy achieve benign cycle and the global ecological security be guaranteed successfully.

Standing at a new historical juncture, we need to move faster in developing the system of ecological conservation. By 2035 when the environment health is fundamentally achieved, the goal of building a beautiful China will be realized basically. By the middle of the twenty-first century, China will have completed in an all-round way the building of ecological, political, spiritual and social civilizations as well as the beautiful China, where people pursue a green development pattern and lifestyle. The successful practice of China's building of ecological civilization is not only a positive contribution to the global ecological security but also a China's plan for promoting the transformation of ecological civilization globally.

6. CONCLUSION

The scientific assertion that lucid waters and lush mountains are invaluable assets has been proved constantly in the course of reform and opening-up to be theoretically significant and practically measurable. In the initial stage of reform and opening-up, nature was directly used as high-quality assets in exchange for wealth. In the second period (i.e., the period of rapid industrialization and urbanization), equal importance was attached to both development and environment. People began to know that nature itself was wealth. Actions were taken to protect the rivers and mountains from pollution. Then, at the beginning of the twenty-first century, China shifted its focus onto heavy chemical industry, with its status as the world factory being continuously strengthened. China's "lucid waters and lush mountains" initiative contributed to the development of the whole world. On the one hand, the trend of green development worldwide also promoted China's green economic and social transformation. The concept of ecological conservation grew from theory to practice, with environmental degradation being restrained and pollution on the fall. In the 2010s, when China's economic development entered a new normal (which means slower development), it also marked the ecological civilization's entry into a new era. Environmental liabilities are being paid off step by step, and the value of ecological assets has been maintained and increased. The initiative that lucid waters and lush mountains are invaluable assets signifies a new realm of understanding and practice, where man and nature coexist harmoniously in a beautiful China, contributing to global ecological security.

Notes

1 Synchronous development means the synchronous planning, implementation and development of economy, urban and rural construction and environmental protection, so as to realize the unity in economic, social and environmental benefits.
2 The Three Major Policies refer to "prevention first; prevention and control integrated," "whoever pollutes is accountable for remediation" and "strengthening environmental management." The eight major management systems include "environmental impact assessment," "the synchronous planning, implementation and development," "charge of pollution discharge," "responsibility for environmental protection objectives," "quantitative assessment of comprehensive urban environmental remediation," "registration and permit for pollution discharge application," "remediation within time limit" and "centralized control."
3 Li Zhiqing. (2018). Forty Years of Environmental Protection from the Perspective of Economic Development. *Prosecutorial View*, No. 13.
4 Ma Weihui. (2018). Four Decades of Environmental Protection in the Eyes of Qu Geping. *Huaxia Daily*, July 30.
5 The fifteen kinds of polluting small enterprises refer to small factories engaged in the production of paper, leather and dyes and that of, with backward technologies, coke, sulphur, arsenic, mercury, lead, zinc, oil, gold, pesticide, bleaching and dye, electroplating, asbestos products and radioactive substance.
6 Ma Weihui. (2018). Four Decades of Environmental Protection in the Eyes of Qu Geping. *Huaxia Daily*, July 30.
7 Zhuang Guiyang, Bo Fan and Zhang Jing. (2018). China's Role and Strategic Choice in Global Climate Governance. *World Economics and Politics*, No. 4.
8 Zhou Hongchun. (2018). New Ideas, New Tasks and New Measures of Ecological Environment Protection in China. *China Development Observation*, No. 6.
9 Zhuang Guiyang, Bo Fan and Zhang Jing. (2018). China's Role and Strategic Choice in Global Climate Governance. *World Economics and Politics*, No. 4.
10 Qi Qiaoling. (2017). What Wisdom Does the Concept of "Lucid Waters and Lush Mountains Are Invaluable Assets" Generate in Practice?--An Overview Huzhou Conference, *China Ecological Civilization*, No. 6.

CHAPTER TWO

Environmental Policies: Evolution and Evaluation

ZHOU HONGCHUN[*]

1. INTRODUCTION

The environment is an integral part of the natural system, the product of the interaction between man and nature, and the precondition and outcome of economic growth. Environmental problems exist throughout the whole process of economic and social development. The environmental policies, covering laws, regulations, economic policies, environmental standards, supervision and implementation, are the systematic design for the solution to environmental problems. The CPC's environmental claims, decrees, measures and methods can be generalized as environmental protection and economic development going hand in hand.

Over the past forty years of reform and opening-up, China's policies on environmental protection have been not only a model use of international experience but also a constant adjustment of slogans, concepts and law enforcement efforts, which is largely based on the country's basic conditions and the characteristics of

---

[*] Zhou Hongchun: PhD, former Deputy Inspector and researcher of the Development Research Center of the State Council, scholar entitled to the special government allowance of the State Council. His research fields cover resources, environment and industries, and policies related to sustainable development. He published ten monographs, such as *The Circular Economy*, *Low-Carbon Economics*, *Introduction to Economics of Green Development*, among more than 500 research papers.

development stages. After meeting people's basic needs for food, clothing, shelters and vehicles, China has toughened its law enforcement on environmental violations to further satisfy public demands for better living surroundings. This chapter summarizes the evolution and characteristics of China's environmental protection policies in the four decades of reform and opening-up, with a brief assessment and analysis presented. At the end of this chapter are some expectations and policy suggestions.

2. THE DEVELOPMENT OF ENVIRONMENTAL PROTECTION POLICIES AND THEIR KEY POINTS

Environmental protection is the first area in which China tried to adopt international standards, with the government's environmental agencies taking the lead in introducing and advocating for such notions as clean production, sustainable development and circular economy. The fact of lagging public awareness and understanding of environmental protection is due largely to the stage of development of China. Reflecting China's distinct approach to addressing environmental problems, one that is characterized by the willingness to learn from others in the world, strong government leadership and meticulous planning, the country's battle against pollution moved progressively from ideas to polices and then to concrete actions. Today, the country as a whole takes environmental protection very seriously, which should pave the way for sustained improvement in our environmental conditions.

2.1. The enlightenment and initial stage of environmental protection in China started before the reform and opening-up

In the first few years since the founding of the People's Republic of China in 1949, environmental protection was not a prominent problem, as the government took the restoration of production and economic construction as its main tasks at that time. During the First *Five-Year Plan* period (1953–1957), the priority was given to heavy industry, but equal importance was also attached to environmental protection. For example, all the 156 large- and medium-sized projects were equipped with facilities for sewage treatment and smoke and dust removal. In 1956, the Ministry of Health and the National Construction Committee jointly issued the *Hygienic Standards for the Design of Industrial Enterprises*, laying down requirements for environmental protection.[1]

China became highly concerned about environmental protection due largely to the pollution accidents at home and the environmental movements abroad.

The year 1972 saw accidents such as the pollution in Dalian Bay, the polluted fish in Beijing and the water pollution in Songhua River system, among many other incidents. But few Chinese people knew what environmental protection and pollution nuisance meant, nor did they know that environmental protests were surging internationally and that Western countries had embarked on a large-scale environmental protection movement since the 1960s marked by the publication of Rachel Carson's *Silent Spring*. Also, in 1970, the US launched an "Earth Day" campaign, the theme of which was to protect the environment. On June 5, 1972, directed by Premier Zhou Enlai, the Chinese delegation participated in the United Nations Conference on Human Environment held in Stockholm, Sweden. This was the first large-scale international conference that China attended after the resumption of its UN seat. At the conference, Chinese delegation put forward the views that the sovereignty of each country be respected and each country be treated equally regardless of their sizes, which was well received by the developing countries and had contributed to the revision of the *Declaration on the Human Environment*.[2]

At this point, China's environmental protection entered the stage known as the enlightenment and initiation, mainly marked by the following events:

> First, the government proposed the system of the synchronous planning, implementation and development of economy, urban and rural construction, and environmental protection, which was also known as the "three-synchronous system" in China. In June 1972, the State Council approved the *Report of the State Development Planning Commission and the National Construction Commission on the Pollution and Solutions of Guanting Reservoir*, which was the first time the "three-synchronous system" was put forward. Between August 5 to 20, 1973, the first National Environmental Protection Conference was held in Beijing, laying down the "Thirty-Two-Character Principles" of "overall planning and rational distribution; utilizing resources comprehensively and turning hazards into benefits; relying on the people and involving everybody; and protecting environments for the benefit of the people." The *Several Provisions on Protection and Improvement of the Environment* was adopted. The State Council endorsed the document as State Council Doc [1973] No. 158.
>
> Second, the government began to issue regulatory standards. The *Opinions on Further Promoting the Dust Removal from Chimneys* was promulgated in April 1973, along with the *Three Industrial Wastes Emission Standard (Trial)* (GBJ 4–73) released in November of the same year. What followed included the *Interim Provisions on Prevention of Pollution in Coastal Waters* (1974), the *Provisions on Radiation Protection (Internal Trial)* (1974), the *Sanitary Standard for Drinking Water (Trial)* (1976) and the *Several Regulations on*

Comprehensive Utilization for Treatment of Three Industrial Wastes (1977). In February 1978, the Constitution of the People's Republic of China was promulgated, stipulating that "the state protects the environment and natural resources, and prevents pollution and other public nuisance."

Third, the government started to carry out environmental protection planning. In 1975, the Environmental Protection Leading Group of the State Council required all regions and departments to include environmental protection into their long-term planning and annual plans. In 1976, the *Circular on the Preparation of the Long-Term Planning for Environmental Protection* stipulated that environmental protection be incorporated into the long-term and annual plans of the national economy.[3]

Fourth, environmental protection agencies were set up one after another. In October 1974, the Environmental Protection Leading Group of the State Council was established, with an office to handle daily work. According to the relevant provisions of Document No.158 of the State Council, competent authorities in all regions successively established environmental protection institutions.

In short, China's environmental conditions were good as a whole before the reform and opening-up, despite some local and point pollution accidents that happened from time to time. The idea that "environmental pollution is a matter of capitalism while the problem of food supply is what socialism should solve" was partly accepted.[4] With his foresight and keen awareness of the seriousness and urgency of the environmental problems, Premier Zhou Enlai helped China to embark on the journey of environmental protection.[5]

2.2. China started to develop environmental protection ideas as pollution began to occur (1979–1991)

After the Third Plenary Session of the Eleventh CPC Central Committee held in December 1978, China entered the initial stage of the establishment of its legal and regulatory system regarding environmental protection, at the juncture of the transformation of the country's development strategies. The major and representative events or activities that took place in this initial period were as follows.

First, environmental protection was identified as a basic state policy. The second National Environmental Protection Conference was held in Beijing from December 31, 1983, to January 7, 1984. At the conference, Li Peng, then Vice Premier, announced on behalf of the state council that environmental protection was a strategic mission in China's modernization drive and a basic state policy, which laid a solid foundation for China's environmental efforts in the future.

Second, China formed a framework comprising three major policies and eight major management systems, which were put forward at the third National Environmental Protection Conference held at the end of April 1989. The three major policies refer to "prevention first; prevention and control integrated," "whoever pollutes is accountable for remediation" and "strengthening environmental management." The eight major management systems include "environmental impact assessment," "the synchronous planning, implementation and development," "charge of pollution discharge," "responsibility for environmental protection objectives," "quantitative assessment of comprehensive urban environmental remediation," "registration and permit for pollution discharge application," "remediation within time limit" and "centralized control."[6] The eight systems could be further classified into the old three systems (the first three systems) and the five new systems (the latter five systems).

Third, environmental protection was incorporated into national economic and social development plans. In 1982, the *National Economic Plan* was renamed as the *National Economic and Social Development Plan*. The *Sixth Five-Year Plan* had environmental protection as an independent chapter, which included six environmental indicators, major tasks and main countermeasures. China took the lead in using national economic plans to promote environmental integrity.

Fourth, laws and regulations grew into a system. In 1979, China promulgated *Environmental Protection Law (Trial)*. By the end of 1991, China promulgated a total of 12 laws on resources and environment, more than 20 administrative regulations, over 20 departmental regulations, 127 local laws, 733 local regulations and a large number of normative documents. The legal system for environmental protection had been preliminarily established, which laid a solid legal foundation for strengthening environmental management.[7]

Fifth, the Environmental Protection Bureau was upgraded to a unit directly under the State Council. In 1982, the Ministry of Construction was renamed the Ministry of Urban and Rural Construction and Environmental Protection, with Environmental Protection Bureau as its subsidiary. The Environmental Protection Commission was established under the State Council in 1984 at the end of which the Environmental Protection Bureau under the Ministry of Urban and Rural Development and Environmental Protection was reorganized into National Environmental Protection Agency. In 1988, the environmental protection responsibilities were severed from the Ministry of Urban and Rural Development and Environmental Protection to the newly founded National Environmental Protection Agency (a sub-ministerial level agency), which was defined as the competent department directly under the State Council.

Sixth, environmental education started. In 1980, the *Environmental Education Development Plan (Draft)* was issued, requiring that environmental protection be incorporated into the national education plan. In 1981, the

Environmental Management Cadres College (now renamed as Hebei University of Environmental Engineering) was established in Qinhuangdao City. According to the *Decision of the State Council on Further Strengthening Environmental Protection* released in 1990, educational and publicity departments should include environmental publicity and education in their plans. In 1991, the State Education Commission classified environmental science as a first-level discipline.[8]

At this stage, environmental pollution began to appear in China, with the water quality deterioration of urban rivers being the most typical. Environmental losses accounted for 10% to 17% of GNP. The discharge of some pollutants exceeded the environmental carrying capacity. Take the total discharge by China in 1991, which stood at 16.22 million tons (excluding the discharge by township enterprises) as an example. It was higher than the maximum ecological carrying capacity of China then, which was 16.2 million tons.[9]

2.3. Worsening environmental pollution urged faster legislation (1992–2002)

After Deng Xiaoping delivered his important speeches on his tour to southern China in 1992, the Central Committee of the Communist Party of China and the State Council established the strategy of rejuvenating the country by science and education as well as the strategy of sustainable development. Since then, China began to put into place an environmental protection system with the control over total discharge as the core.

First, environmental protection became a topic of the National Symposium on Population. During the "Two Sessions" (the annual sessions of the National People's Congress and the Chinese People's Political Consultative Conference) held in 1991, President Jiang Zemin suggested to convene a "National Symposium on Family Planning," which was renamed "National Symposium on Family Planning and Environment" in 1997 and "National Symposium on Population, Resources and Environment" in 1999. Jiang Zemin pointed out at the symposium convoked in March 1999 that the importance of handling the relationship between economic construction and population, resources and environment be deeply understood from a strategic point of view, which would help to establish the focus of environmental protection at different stages and is also of great significance for local Party committees and governments to promote environmental protection.

Second, the environmental protection strategy changed. The UN Conference on Environment and Development held on June 12, 1992, led to China's *Ten Countermeasures for Environment and Development*. In October 1993, the second National Symposium on Industrial Pollution Prevention and Control summarized the experience and lessons drawn from industrial pollution prevention and control and proposed to make three changes, namely, to change from terminal control to the control of the whole production process, from pollutant concentration control to the control of both pollutant concentration and total discharge, from dispersed control to the combination of dispersed and centralized control. The year 1994 witnessed the proclamation of *China's Agenda 21*, which is also known as the *White Paper on China's Population, Environment and Development in the 21st Century*, an overall strategy, plan and countermeasure scheme for sustainable development drafted by China in accordance with the *Agenda 21* adopted by the United Nations at the UN Summit on Environment and Development in 1992; thus, it serves as a guiding document for the Chinese government to draft the mid- and long-term social and economic development plans.

Third, the legislation for environment protection was accelerated. In 1993, the National People's Congress established the Environmental Protection Committee, which was later renamed the Environment and Resources Protection Committee. They formulated and promulgated five laws, including the *Law on Promoting Clean Production* and the *Law on Environment Impact Assessment*, in addition to the revision of other three laws such as the *Law on the Prevention and Control of Atmospheric Pollution* and the *Law on Prevention and Control of Water Pollution*. The State Council drafted or revised over twenty regulations with the *Regulations on Nature Reserves* included, laying down more than 200 environmental standards.

Fourth, China sought to create a more integrated policy system from economic and technological perspective for environmental protection. The government adopted policies with regard to industries' investment, finance and tax, price, import and export to ensure that enterprises saving and comprehensively utilizing resources could benefit and those violating environmental regulations be punished. In 2002, China formulated the *National Policy on Industrial Technology*, which stipulated that the emphasis be placed on the development of high technologies and industrialization so as to reform and upgrade traditional industries through advanced technologies, specifying development directions of new-energy technology, energy and environmental protection, raw materials, construction industry and so forth.

Fifth, the reform of environmental impact assessment system was carried out. The government began to implement the classified management of environmental impact assessment for development and construction projects by introducing competition mechanism, the pilot bidding system, the pilot post-assessment and the accountability mechanism. The *Administrative Measures for Regional Environmental Impact Assessment of Development Zones* was issued in order to control the discharge of pollutants from the source.

Sixth, the pilot pollutant discharge permit system was rolled out. Since 1992, major cities throughout the country began issuing water pollutants discharge permits. The National Environmental Protection Agency issued *Several Opinions on Further Promoting the Pilot Work of the Permit System for Atmospheric Pollutants Discharge, Principles and Methods for Determining the Emission Indicators for Atmospheric Pollutants Discharge Permit, Administrative Measures for Atmospheric Pollutants Discharge Permit* and so forth, selecting Taiyuan, Liuzhou, Guiyang, Pingdingshan, Kaiyuan and Baotou to carry out pilot emission trading.

Seventh, China commenced to promote the environmental labelling system. *The Outline of China Environmental Protection Work (1993—1998)* called for the establishment and implementation of environmental labelling system, which was an important move toward the application of the ISO14000 family of environmental management standards developed by International Organization for Standardization (ISO) in 1993. As a result, China Certification Committee for Environmental Labelling Products was established on May 17, 1994.

Eighth, China implemented the total volume control and the cross-century green projects. The fourth National Environmental Protection Conference convened in July 1996 and the *Decision of the State Council on Strengthening Environmental Protection* put forward the objectives of "One Control, Double Aims" ("One control" means the control of the total amount of pollutants should be within the total amount specified by the State by the end of 2000; "Double aims" mean first that the industrial pollution source shall meet the national or local standards for pollutant discharge and second that air and surface water shall meet national environmental quality standards set according to functional areas) and "332 Projects," which included the treatment of three lakes (Taihu Lake, Chaohu Lake, Dianchi Lake), three rivers (Huaihe River, Haihe River, Liaohe River) and two areas (areas of acid rain and sulfur dioxide pollution control). 15 categories of small enterprises such as small coal mines, small coking factories and small paper mills, which were of high resource consumption, serious pollutant discharge and non-compliance with industrial policy requirements, were forced to close. To realize these objectives, we must toughen the management, promote proactively the transformation of economic growth mode, gradually increase environmental investment and strengthen the rule of law in environmental areas, namely Four Musts.

Ninth, China supported clean production and environmental protection industry. As part of an array of efforts to encourage clean production since 1993, China promulgated *Several Opinions on the Promotion of Clean Production* (1997), the *Circular on Improving the Development of Environmental Protection Industry, Current Catalogue of Key Industries, Products and Technologies the Development of*

Which Is Encouraged by the State (Revised in 2000), *Catalogue of Environmental Protection Industry Equipment (Products) Encouraged by the State, the Opinions on Accelerating the Development of Environmental Protection Industry*, requiring that environmental protection industries be given preferential policies regarding tax reduction and exemption.[10]

Tenth, China strengthened its environmental education. In 1992, China put forward the idea that "education is the basis of environmental protection." After years of development, environmental education now covers basic education, professional education, adult education and social education, on top of being an integral part of the curriculum of primary and secondary schools. The institutions of higher learning set up the specialty of environmental engineering with a complete system of courses and a reasonable structure, which can basically meet the needs of environmental protection undertakings for talents.

Eleventh, National Environmental Protection Agency was upgraded and the China Council for International Cooperation on Environment and Development (CCICED) was established. In 1998, the State Council promoted the National Environmental Protection Agency to State Environmental Protection Administration while revoking the Environmental Protection Commission. The year 1992 saw the establishment of CCICED, which aimed to absorb international suggestions on economic and environmental development, on the one hand, and reach out to the world community on China's environmental policies and attitudes, on the other hand.

After Deng Xiaoping's visits to southern China in 1992, the country ushered in a new round of economic boom. But as a Chinese proverb goes, "turnips are not washed when selling fast," which means, in this case, that environmental health was compromised for economic gains. This situation was complicated by the brisk and disorderly development of township enterprises, giving rise to graver environmental pollution. For example, some rivers and lakes turned black and malodorous, which, along with blue algae eruption, jeopardized drinking water safety. Smog enshrouded many cities, directly causing the outburst of respiratory diseases among urban residents. To make it worse, pollution, which began to expand from cities to the countryside, from east China to western regions, not only led to the increasing number of environmental accidents but also grew into the hotspots of social complaints. Since 1995, surveys on public awareness of environmental protection, which were conducted by China Environmental Protection Foundation and the National Survey Research Center at Renmin University of China (NSRC), showed that Chinese people had a relatively low level of environmental knowledge.[11] This vividly demonstrated the public environmental understanding at that time.

2.4. The number of pollution-induced mass incidents increased with environmental policies gradually improving (2003–2012)

When the wheels of history rolled into the twenty-first century, the CPCCC and the State Council put forward the scientific outlook on development, which puts people first and aims at comprehensive, coordinated and sustainable development. On the National Symposium on Population, Resources and Environment held in 2003, Hu Jintao, then General Secretary of the CPC Central Committee, stressed that environment protection should ensure public access to clean water and air and safe food so that people could work and live in a good environment, and that top priority could be given to addressing problems that posed the most serious threats to public health. The national symposiums on population, resources and environment till 2005 set the people's livelihood as a target of environmental protection.[12]

First, China issued policies related to industries, prices and foreign trade. These policies included but were not limited to the *Notice of the General Office of the State Council on Distributing the Opinions of the National Development and Reform Commission and other Departments on Clearing and Rectifying Calcium Carbide and Ferroalloy Industries*, the *Emergency Notice on the Several Opinions on Clearing and Standardizing the Coke Industry* (NDRC Industry [2004] No. 941) and the *Notice on Implementing the Opinions on Further Strengthening the Outcome of Clearing Calcium Carbide, Ferroalloy and Coke Industries and Regulating their Healthy Development* (NDRC Industry [2004] No. 2930). All these documents intended to standardize or limit the development of industries notorious for high-energy consumption and pollution so as to optimize and upgrade industrial structure.[13] Besides, China adopted preferential tax policies for enterprises in environmental protection industry, such as an income tax policy known as the "three years of tax exemptions and three years of tax-halvings," which meant first, the exemption or immediate refund of value-added tax on enterprises in sewage, reclaimed water and waste disposal industries; second, the reduction of value-added tax on desulfurized products by half; and third, the exemption corporate income tax for investment in purchasing environmental protection equipment. All these measures played a positive role in pollution abatement. In addition, China implemented price policies for environmental protection and opened a new path for the market-oriented development of air pollution control industry. The year 2004 saw the introduction of a policy of giving RMB0.015 per kilowatt hour to electricity generated with desulfurization technology, which boosted the rapid growth of desulfurization-related industries of power plants. The proportion of the installed capacity of desulfurization units in that of thermal power plants increased rapidly from 8.8% in 2004 to 87.6% in 2011, when the government further formulated the policy of giving another RMB0.008 per kilowatt hour to electricity generated with desulfurization technology as well as on-grid electricity price incentive for power generation through waste incineration. What is more, China adjusted tax rebate policy for import and export. For example, the Ministry of Finance and the State Administration of Taxation adjusted the export tax rebate policy by issuing, in 2004, an urgent notice on stopping the export tax rebate of coke and coking coal. In 2005, seven departments, including the NDRC, jointly released a document which played a positive role in curbing the export of products from the industries of high pollution, intensive energy and resources consumption. The phenomenon of deforestation has been remarkably reduced.

Second, the government expanded financing channels for environmental industries. To begin with, China listed resources conservation, circular economy and environmental protection and so forth as the investment priorities of government bonds. In 2004, the Ministry of Finance issued the *Administrative Measures for the Use of Central Subsidy Special Funds for Local Clean Production* to support the cleaner production by small and medium-sized enterprises in petrochemical, metallurgical, chemical and paper industries. The national bonds investment was guided to the construction of biogas projects in rural areas, the continued construction of supporting facilities to provide areas with irrigation and drainage, as well as the water-saving re-engineering projects. The national bonds investment supported a number of technological development and industrialization projects such as the technologies of saving and replacement, energy cascade utilization, recycling and "zero emission." On top of the first measure, the government also carried out the preferential financing policy for environmental protection. Let us take the "green credit" policy, which has been implemented since July 2007, as an example. By the end of 2010, the balance of green credit of China Development Bank and the four major state-owned banks, namely, BOC, ICBC, CBC and ABC, reached RMB1,450.6 billion. During the Tenth *Five-Year Plan* period, environmental protection loans amounted to RMB118.3 billion, accounting for 14% of the total national investment in environmental protection during the same period. During the Eleventh Five-Year Plan period, energy conservation and emission reduction sectors received a loan of RMB586 billion, of which over 320 billion was directed to environmental protection, accounting for 15% of the total amount of environmental protection investment in the country during the same period. In addition, some environmental protection companies partly solved the problem of pollution control funds through their IPO in Hong Kong SAR of China, the US, Germany, Japan and so on.[14]

Third, China promoted the franchise system. In 2002, the marketization reform marked by the franchise system started, changing the situation where sewage and waste disposal plants were mainly built by government and the environmental protection facilities, were too often laid idle, which generated little environmental benefits. The construction of environmental protection facilities was accelerated when private capital and foreign capital entered the fields of water, gas and heat supply, sewage and waste disposal.

Fourth, the public got involved in planning for environmental assessment. On April 13, 2005, State Environmental Protection Administration held the first hearing on the implementation of the *Law on Appraising of Environment Impacts*. In March 2006, the *Interim Measures for Public Participation in Environmental Impact Assessment*, a normative document on public participation in environmental protection was issued as a "control gate" to prevent pollution and ecological destruction.

Fifth, China advanced the development of eco-industrial parks and circular economy. In 2003, the State Environmental Protection Administration issued the *Guidelines for the Planning of Circular Economy Demonstration Zone (for Trial Implementation)* to promote the recycling of materials, cascade utilization of energy and the reclamation of waste. The *Opinions of the State Council on Accelerating the Development of Circular Economy* released in 2005 became the guiding document for the development of circular economy.

Sixth, China dispatched watchdogs to strengthen law enforcement. In February 2006, the State Environmental Protection Administration and the Ministry of Supervision promulgated the *Interim Provisions on the Punishment of Violations of Environmental Protection Law and Disciplines*. From 2003 to 2006, seven departments, including the State Environmental Protection Administration, launched a special campaign to rectify the polluters and safeguard public health. On January 18, 2005, the State Environmental Protection Administration started an "environmental protection storm" by suspending or denying the review and approval of projects that failed to meet the environmental standards. In 2006, State Environmental Protection Administration (SEPA) set up eleven local law enforcement supervision agencies.[15] The construction of radioactive waste repository and database began, with radioactive sources being under primary control. In January 2007, the review and approval of radioactive construction projects were suspended or restricted, or "restricted in certain regions."[16] A year later, the SEPA was upgraded to the Ministry of Environmental Protection, becoming a ministerial level agency under the State Council, which played an important role in the comprehensive decision-making for socioeconomic development and environmental protection.

Seventh, the Trans-Century Cross-China Environmental Protection Tour was launched and became an important channel for environmental information disclosure. Since 1993, the Trans-Century Cross-China Environmental Protection Tour, which was headed by the Environmental Protection and Resources Conservation Committee of the National People's Congress and joined by another fourteen ministries and departments including the CPCCC Publicity Department and the National Environmental Protection Agency, helped to solve major environmental problems such as the cutoff of Yellow River, the pollution of Huaihe River and Bohai Sea, the pollution in Shanxi Province, Shaanxi Province and Inner Mongolia (which was known as the Black Triangle), and the over-hunting of wild animals.[17] This initiative opened up a new way to supervise environmental pollution China-wide according to law and through public opinions. Moreover, NGOs emerged as an important player in environmental protection. According to the *Report on the Development of Environmental Protection NGOs in China*, by the end of 2007 there were 2,768 environmental protection NGOs, of which 1,382 were sponsored by government, accounting for 49.9%; 1,116 by student associations and unions, taking up 40.3%; 202 of spontaneous organization, standing at 7.3%; and 68 by the branches of the international environmental protection NGOs, amounting to 2.5%. These environmental protection NGOs were mainly located in Beijing, Tianjin, Shanghai and the eastern coastal areas.[18] However, due to their small scale and poor working conditions, environmental protection organizations had not fully exerted their influence, especially when it came to the fact that 60% of them even did not have independent offices.

Just as a proverb goes, every coin has two sides. Since the second half of 2002, China witnessed a new round of high-speed growth when many cities set up projects of high-energy consumption and high emission, such as steel, cement, chemical industry and coal power. Despite the fact that China began to implement the plan of energy conservation and emission reduction, the momentum behind the expansion of heavy and chemical industry, just like the rising trend of pollutant emission, was too hard to contain. In 2006, the amount of major pollutants emitted in China was as follows: 25.88 million tons of sulfur dioxide, 15.23 million tons of nitrogen oxide, 14.28 million tons of chemical oxygen demand (COD) and 1.41 million tons of ammonia nitrogen. Since then, the government increased its efforts to reduce emissions and supplemented it with market-based measures, leading to the gradual reduction of major pollutants. The environmental quality, however, did not see improvement alongside. Instead, pollution accidents continued happening one after another, resulting in frequent public incidents.[19]

2.5. Environmental protection has risen to the strategic height in ecological conservation since the Eighteenth CPC National Congress

Since the Eighteenth National Congress of the Communist Party of China held in 2012, the whole country has been more conscientious and proactive in promoting the concept of green development, with the neglect of environmental protection being reversed and the building of ecological civilization seeing tangible results. This is a concrete embodiment of the spirit of CPCCC with Xi Jinping as the core being highly responsible for the sustainable development of the country and the nation, as well as a positive response to public demands.

First, China promulgated environmental laws, regulations and policies more frequently than ever before. Since the Eighteenth CPC National Congress, the CPCCC with Comrade Xi Jinping at its core promoted much fundamental, long-term and pioneering efforts, with the Central Leading Group for Comprehensively Deepening Reform reviewing and approving more than forty reform plans on ecological civilization and environmental protection. The State Council issued the *Action Plan on Prevention and Control of Air Pollution* and the *Action Plan on Prevention and Control of Water Pollution* in 2013 and 2015, respectively, and the *Action Plan on Prevention and Control of Soil Pollution* in 2016. The *Environmental Protection Law*, which came into force in 2015, is known as the strictest one of its kind in history, playing a positive role in improving environmental quality.

Second, the central government adopted a great variety of measures for the inspection and supervision of environmental protection, which have achieved remarkable results. China has very comprehensive environmental policies and regulations, the implementation of which, however, relied upon special environmental protection campaigns previously. Unlike in the past, the current environmental inspection and supervision by central government has introduced diverse law enforcement measures with much harsher corresponding penalty. These measures include the inspection and supervision of enterprises and that of government, with the former strengthening examination and surveillance and the latter involving general environmental protection inspection and special-case supervision, along with other means such as the appointed conversation, restricted review and approval, notification of cases, and the listed supervision by governments of higher level. Besides, the judicial means have also been well applied. In the use of criminal and civil forces, the Ministry of Public Security, the Supreme People's Court and the Supreme People's Procuratorate carried out close cooperation. In 2017, the cases filed by the environmental protection agencies at public security departments fell into two categories: administrative detention and suspected environmental crimes, which totaled over 8,600 and 2,700, respectively, with the former increasing by 112.9% and the latter by 35% over 2016. The Ministry of Environmental Protection also supported public-interest organizations in filing five environmental civil public interest litigations. When it comes to the means of law enforcement, remote sensing, online monitoring and Big Data analysis are used among many others. Grid technology is also employed in monitoring air quality in Beijing, Tianjin and 26 cities in Hebei, Shanxi and Henan, where cities are divided into grids of 3x3 km^2 which are equipped with monitoring devices. Where there is an abnormally high concentration of pollutants detected, law enforcers will go immediately to the site to inspect the situation. The impact of the supervision conducted by the Central government is unprecedentedly huge. The inspection and supervision by central government and that of special cases have been well acknowledged by the public, the CPCCC and local govrnments, as they did solve the problems. By the end of 2017, central government had dispatched four batches of inspectors, who helped to address many outstanding environmental issues, holding accountable over 18,000 people. In the first batch, those who were found guilty mainly came from eight provinces totaled 1,100, with 130 of them holding official ranks of director or deputy director of departments and bureaus at provincial level while in the second batch, the number stood at 1,048, with three of them holding officials ranks of governor or minister and fifty-six of them holding official ranks of director or deputy director of departments and bureaus at provincial level. The attitude of the Ministry of Environmental Protection was clear: any accident, wherever it is found, is to be strictly investigated; and those who contribute nothing in everyday work and have misconducts in law enforcement shall be resolutely opposed, allowing no such misdeeds to create chaos in the inspection and supervision by the central government in general.[20]

3. A COMMENT ON THE CHARACTERISTICS OF EVOLUTION OF ENVIRONMENTAL PROTECTION POLICIES

The road of China's environmental protection is much like "crossing high mountains." The environmental problems that had occurred in different stages of the century-long industrialization of developed countries emerged in China all at the same time, displaying characteristics of being structured, compound and condensed. If the ecological destruction is due largely to the overdevelopment of agriculture in history, the environmental degradation is the direct result of industrialization.

3.1. Environmental policies evolved from slogans and ideas to action taken by "Government + Market"

The concept of environmental protection in our country, in its early stage, was mainly of slogan and political mobilization, with technology "salon" being the typical attribute of competent authorities. The turning point appeared in 1998 when China stipulated in the *Law of Land Administration* that whereas reclamation of a land or rounding up of a land for reclamation would give harm to ecological environment the land concerned should be restored as forests, pasture fields or lakes. Chinese environmental policies, laws and regulations keep improving, matching their foreign counterparts almost in every respect. *China's Agenda Twenty-One*, the Trans-Century Cross-China Environmental Protection Tour and the inspection and supervision by central government were all of political mobilization in essence, aiming at raising the environmental awareness of local Party and government officials as well as enterprises.

The environmental laws and regulations in their early stage were not easily enforced, leading to low levels of compliance. In many cases, lawbreakers were not prosecuted. In some places, the environmental regulations and standards were not strictly implemented, with some people taking advantage of loopholes. Just as a Chinese proverb goes, "thunder is loud but rain is light," which means that the lawbreakers would be punished severely but few actions were taken. Some local officials paid scant attention to environmental protection, only tightening it when inspected, which is vividly described by many as "one eye open and the other closed." The low cost of law-breaking and high cost of law-abiding caused a situation where polluters competed against each other in asking for preferential policies and funds from government for pollution treatment, forgetting that meeting the pollutant discharge standard was their innate responsibility.[21] When an enterprise spent millions of dollars or more to abate the pollution but the fines for law-breaking cost 90% less, it would not be possible to generate a deterrent effect but to produce a negative "demonstration effect" instead, resulting in "adverse selection." Companies would rather be fined than treat pollutants. What made it worse was that the government failed to take into full consideration the cost of resources and environment when they tried to attract investment for the sake of GDP. The result is that many incoming investors were heavy polluters. In consideration of the tax revenue these polluting enterprises contributed, local governments often chose to close both eyes.

The Third Plenary Session of the Eighteenth Central Committee of the CPC issued the *"Decision of the CPC Central Committee on Major Issues Concerning Comprehensively Deepening Reform"* requesting to establish and implement the lifelong accountability system for environmental pollutions, which played an important part in the decision-making by governments at all levels. After environmental protection became a national policy, government officials became the most decisive factor regarding its enforceability and environmental performance. In the past, relevant persons in charge in local governments would be held accountable and punished for any environmental accidents that happened. Now, the *Environmental Protection Law* specifies the principle of territorial management, according to which leaders of local Party and governments bear the overall environmental responsibility. That means to start with the "key few" decision-makers, the top leaders in particular, and make them fully aware of the importance of ecological conservation. The government also promoted the establishment of a green development model based on environmental carrying capacity, for which a monitoring and early warning mechanism was set up along with the environmental accountability system and performance assessment system. The impact on the environment was taken into consideration when drafting economic policies, with restrictions on the review and approval of environmental projects from areas that failed to meet environmental targets, in an attempt to better environment quality step by step and lay a foundation for the environmental legislation.

Environmental protection is a typical externality problem in economics, which requires the government to play its due role in making up for "market failure." Practice has proven that environmental remedies, however good, will be difficult to implement without the commitment of government officials, especially the "top leaders." Therefore, changing development modes and adjusting structures have also become vital actions China took for environmental protection. By changing the previous practice of "beggar-thy-neighbor" in pollutant emissions, our country made it clear that the environmental protection objectives should be incorporated into the economic and social development planning and performance assessment indicators for political achievement. In the past, the distorted price of natural resource and the lack of inclusion of environmental cost were largely responsible for environmental pollution and ecosystem degradation. Accordingly, China issued the policy of internalizing the externality cost of environmental protection, with the emission permit system being one of them, in order to implement the system that whoever pollutes the environment and damages the ecology must pay compensation, which integrates the pollution management expenses into the cost and makes producers and the consumers bear the responsibility of environmental protection.

The Chinese government has also done much work in introducing market measures in environmental protection and pollution management. Let us take the adoption of policies for environmental economy as an example, which included the trading of emission rights and ecological compensation. China also tried to solve environmental problems by developing environmental protection industry. In Jiangsu Province and Zhejiang Province in the eastern coast of China, BOT and TOT models have been adopted since the beginning of the twenty-first century to treat township garbage and sewage, with remarkable results achieved. China's environmental protection industry, which provides equipment and services for pollution control and emission reduction, pollution treatment and waste disposal, has seen rapid expansion. According to the results of the fourth survey of environmental protection industry, by 2012, there were 230,000 enterprises engaged in environmental protection industry, with a labor force of over 3.19 million employees. As to the three indicators for the development of environmental protection industry, the annual growth rates of environmental products, environmental service industry and resource recycling rate exceeded 30%, 28% and 14%, respectively. From 2004 to 2012, China's environmental protection industry grew by over 20% annually, which was largely attributed to the measures China adopted for energy conservation and environmental protection.

3.2. The functions of the competent environmental authorities are gradually oriented to policy formulation and the supervision of implementation

There have been two outstanding and persistent problems in environment protection in China. First, the responsibilities of government departments overlap, which can be described by a Chinese proverb, "a flooding river is treated by nine dragons," meaning that when something is everybody's responsibility it ends up being nobody's responsibility, for it is not clear who shall shoulder the responsibility for accidents. Second, the government is both the owner and supervisor, or the "player" and "referee" at the same time. Therefore, as a referee, the governments often find their authority and effectiveness clearly insufficient. Still worse, the environmental authorities changed their management strategies during a certain period of time, tightening it only on paper but relaxing it in reality (e.g., the policy of "restricting the review and approval of projects in river basins and specific areas" was no longer implemented after 2006). The appointment of the head of local environmental authorities fell into the power of local government. Therefore, it is not strange that the directors of local environmental protection bureaus are given the task to attract investment but end up in having polluting enterprises instead. In some cities, the local environmental protection bureaus are understaffed and worse still, poorly equipped, which made them incompetent to supervise all enterprises that discharge pollutants.

China's environmental authorities were born from specialized organizations. In 1971, the National Planning Commission set up a leading team for "three wastes" utilization. In 1974, the Leading Team for Environmental Protection was officially established under the State Council. In 1982, the State Council renamed the Ministry of Construction as the Ministry of Urban and Rural Development and Environmental Protection, and Environmental Protection Bureau was one department in it. In 1984, the Environmental Protection Commission was established under the State Council. At the end of 1984, the Environmental Protection Bureau under the Ministry of Urban and Rural Development and Environmental Protection was reshuffled into National Environmental Protection Agency. In 1988, the environmental protection functions were severed from the Ministry of Urban and Rural Development and Environmental Protection to the newly founded National Environmental Protection Agency (a sub-ministerial level agency), which was defined as the competent department under the State Council. In 1998, the National Environmental Protection Agency was upgraded to State Environmental Protection Administration (Ministerial level), as an affiliated institution of the State Council in charge of environmental protection effort. In 2008, the State Environmental Protection Administration was upgraded to Ministry of Environmental Protection, as an integral department of the State Council.

In 2018, the State Council carried out an institutional reform, which created favorable conditions for the governance of the ecological system. The newly founded Ministry of Ecology and Environment integrated all the responsibilities of the Ministry of Environmental Protection with the related responsibilities of other six departments, which would perform unified responsibility for supervising various types of pollution and emissions in both urban and rural areas and the according law enforcement. According to the institutional reform plan of the State Council, the functions of the Ministry of Ecology and Environment mainly include the formulation of environmental policies, regulations and standards, and the monitoring and supervision of environmental violations. This move mirrors the CPC's people-centered development ideology and is a profound change in promoting environment integrity, boosting ecological conservation, modernizing governance system and capacity, and thus a milestone of great realistic and historical significance.

The establishment of the Ministry of Ecology and Environment can help to distinguish the relationship between resource owners and regulators, which are supposed to be independent while cooperating with and supervising each other. It can unify the previously decentralized responsibilities for pollution prevention and ecological protection, and form "five connections," namely, connecting the ground and underground, shore and water, land and sea, urban and rural areas, and the carbon monoxide and carbon dioxide. By unifying the prevention and control of air pollution and the response to climate change, we will create an organizational guarantee for winning the battle against pollution and strengthening the protection of environment.[22]

In short, in terms of environmental protection management, our country has unshackled the thought that environmental protection is the sole responsibility of environmental authorities and has strengthened the obligations that local parties should undertake, which makes the public a major force in environmental protection. Technically, we have shifted from the original "end treatment" to "entire process control." When it comes to economic means, we have transformed from "pollution charge" to the combination of "emission permit" and "ecological compensation." As far as the methodology of governance is concerned, we have evolved from the "point source" treatment to "centralized treatment," making pollution abatement more economical.

3.3. Investment in environmental protection has been continuously increased, with remarkable results achieved in pollution control

Due to the fact that the investment in pollution abatement was insufficient, what we achieved was far less comparable than the negative effect caused by pollutant discharge. The pollution that should be treated remained intact. Some damages took place in the name or disguise of protection. The funds that should be used for environmental protection were misappropriated for other purposes. Some investments did not generate their due effect. For example, some sewage treatment plants failed to operate normally after being built, with some of them even becoming pollution sources. Today, the situation is changing.

Relevant research shows that China's investment in environmental protection reached RMB2.5–3 billion on a yearly basis in the early 1980s, accounting for about 0.51% of GDP; over RMB10 billion by the end of 1980s, about 0.60% of GDP; RMB101 billion in 1995, around 1.02% of GDP; RMB238.8 billion in 2005, or 1.3% of GDP; RMB665.4 billion in 2010, or 1.66% of GDP. Encouraged by the government fiscal investment, social capital gradually becomes the main body of investment. From 1998 to 2002, the government issued treasury bonds of RMB660 billion, including 65 billion for 967 urban environmental facilities projects, which further attracted RMB210 billion local and social capital. In 2008, of the RMB4 trillion investment, 210 billion went to the ecological conservation.[23]

Without capital investment, pollution abatement would not be possible and environmental protection industry would be drained in sources. Since the Eighteenth CPC National Congress, China has increased its investment in the control of major environmental pollutions, in aspects as follows:[24]

(1) Winning the war to defend the blue sky. Since its inception in 2013, the special funds increased from RMB5 billion per year to 16 billion in 2017, mainly supporting air pollution prevention in key regions such as Beijing-Tianjin-Hebei region, the Yangtze River Delta and the Pearl River Delta, with remarkable progress achieved. In 2017, the Ministry of Finance, in conjunction with the Ministry of Environmental Protection and other relevant departments, launched a pilot project on clean central heating in winter in northern China, which supported the renovation of heating supply facilities in 12 pilot cities, including Tianjin and Shijiazhuang, with the operation model of "enterprise-based, government-driven and affordable for residents" initially built.

(2) Air quality has been improving at an unprecedented speed. The State Council issued the Action Plan on Prevention and Control of Air Pollution in 2013, specifying five targets: the average concentration of PM10 to drop by 10% in 338 prefecture-level cities; the average concentration of PM2.5 to drop by 25% in the Beijing-Tianjin-Hebei region, by 20% in the Yangtze River Delta, by 15% in the Pearl River Delta and to sixty micrograms per cubic meter in Beijing. Compared with the data of 2013, PM2.5 dropped by an average of 22.7% in cities at and above the prefecture level in 2017, and by 39.6%, 34.3% and 27.7% in

the Beijing-Tianjin-Hebei region, the Yangtze River Delta and the Pearl River Delta, respectively in the same year. The concentration of PM2.5 has been below 35 micrograms per cubic meter for three consecutive years in the Pearl River Delta, and 58 micrograms per cubic meter in Beijing – two landmark achievements indeed.

(3) Prevention and control of water pollution have been strengthened. In 2015, the central government integrated various resources and established special funds for the prevention and control of water pollution, mainly for the implementation of relevant tasks of the Action Plan on Prevention and Control of Water Pollution. In 2017, the central government allocated RMB11.5 billion to the special fund, which went to areas as follows. First, it promoted the prevention and control of water pollution in key regions and basins, including the Beijing-Tianjin-Hebei region and its surrounding areas, the Yangtze River Delta and the Pearl River Delta, the Yangtze River, the Yellow River, the upper reaches of South-to-North Water Diversion Project, and the Hanjiang River, all of which have been given major support for coordinated governance. Second, it enhanced environmental protection and ecological restoration in the Yangtze River Economic Belt. In accordance with the instruction of General Secretary Xi Jinping on "protecting the Yangtze River Economic Belt instead of developing it on a large scale," funds were allocated from the special funds for prevention and control of water pollution. Third, it was used to conclude the protection projects for lakes of good water quality.

(4) The war to defend lucid water focuses on the treatment of four types of water bodies with focus on "Four Remedial Measures." The first and foremost are the sources of drinking water, which are the people's water tank; the second is black and odorous water bodies; the third refers to water bodies worse than Class V; and the fourth is water that is discharged into rivers, lakes and seas but fails to meet the standards. The first measure is the remedy of industrial parks. Those polluting enterprises were subject to either renovation, relocation or shutdown. Emission standards for key industries have been continuously improved, and deadlines have been set for the failed ones. The second is the remedy of domestic pollution sources, with focus on accelerating the construction of pollution treatment facilities in urban and rural areas and ensuring the normal operation of the constructed ones. The third is the prevention and treatment of non-point source pollution in rural areas, which consisted of two parts, the reduction of pesticide and fertilizer in crop farming, which has achieved zero growth basically, and pollution control for fish breeding and poultry raising, especially the environmental treatment for

large-scale livestock and poultry farming and aquaculture. The fourth is that we listed the protection of the basic flow for the normal operation of rivers' ecosystem as an important agenda. When it comes to the treatment of water pollution in China, we should watch closely pollutant emission reduction on one hand and expand the service function of water ecosystem on the other hand, so as to improve the capacity of water absorption and purification. Since the issuance of the "ten measures" in the Action Plan on Prevention and Control of Water Pollution in April 2015, the proportion of surface water bodies with quality better than class III (GB3838–2002) in 2017 was 6.3 percentage points higher than that in 2012, and the proportion of water bodies worse than Class V decreased by 4.1 percentage points compared with that in 2012. Compared to data in 2015, the proportion of water better than Class III was 1.9 percentage points higher in 2017 while the proportion of water worse than Class V declined from 9.7% to 8.3%, with a decrease of 1.4 percentage points.

(5) Soil Pollution Control and Remediation. In 2016, the central government integrated special funds such as that for the prevention and control of heavy metal pollution to support local governments in implementing the tasks set forth in the Action Plan on Prevention and Control of Soil Pollution. In 2017, the central government allocated RMB6.5 billion according to the factor method to all provinces for the coordinated implementation of various tasks, including soil pollution risk control, monitoring and assessment, supervision and management, and the remediation and treatment of contaminated soil. The funds were also used to support the demonstration of pollution control and a detailed survey of soil contamination in key areas with heavy metal pollution.

(6) Further improving rural environment by promoting governance with award. The central government issued the policy of "promoting governance with awards" for environmental protection in rural areas, with a focus on supporting comprehensive environmental remediation in areas along the South-to-North Water Diversion Project and the execution of tasks specified in the Action Plan on Prevention and Control of Water Pollution. In 2017, the central government allocated special funds of RMB6 billion for the comprehensive improvement of rural environment, increasing support for Shaanxi, Hubei, Henan, Hebei and Shandong provinces which lie along the South-to-North Water Diversion Project for the purpose of pushing these provinces to carry out all-inclusive rural environmental renovation in their counties and cities along the South-to-North Water Diversion Project. The funds were also used to support the renovation of 130,000 established villages specified in the Action

Plan on Prevention and Control of Water Pollution as well as to protect traditional villages.[25]

4. PRIORITIES AND PROSPECTS OF ENVIRONMENTAL POLICIES

Environmental protection is much of public goods, the comprehensive management and regulation of which needs the government to play its due role – to make up for market failure. Therefore, the "battle against pollution" is the duty of governments at all levels. It is advisable to exert "pressure" on polluting enterprises by adopting legal, economic and necessary administrative measures, or using laws, standards and the supervision of law enforcement as the starting point or by means of information disclosure and increasing punishment, so as to alleviate the pressure of economic development on resources and environment.

Environmental abatement is one of the three major tasks of China up to 2020 with focus on urban environment improvement, water environment protection and land degradation treatment. Instead of giving up development for environment, we should solve the resource and environment problems that our country is facing at present by developing environmental protection industry so as to realize the coordination of economic and environmental interaction, and hence creating a virtuous circle.

4.1. A long-term mechanism for pollution control should be formed starting with the treatment of smog

Since 2017, Chinese government has implemented a special plan for addressing the root causes and the treatment of heavily polluted weather, preliminarily established a mechanism in which government, enterprises and society contribute, participate and offer support together. The public plays a critical role in reporting pollution and environmental enforcement. For example, in 2017, the Ministry of Environmental Protection received 170,000 reports from the public via telephone, WeChat and websites, twice as many as the number in 2016 and 3.5 times as in 2014. In the prevention and control of air pollution in Beijing, Tianjin and twenty-six cities in Hebei, Shanxi and Henan provinces, the Ministry of Environmental Protection made public all reports and complaints concerning air pollution at the earliest time while keeping a close eye on 13,000 cases handled by the local authorities until they were completely solved. As a Chinese idiom goes, "we should suit the remedy to case and prepare medicine by filling the prescription." Only in this way can the expected effect of environmental governance be produced.

We have cut down pollutant emissions by optimizing the structures of industries, energy and transportation. Over the past five years, China has made progress in eliminating backward production capacity and reducing the excess capacity in steel and coal industries. In the *2018 Report on the Work of the Government*, Premier Li Keqiang pointed out that steel production capacity declined by more than 170 million tons and coal production capacity dropped by over 800 million tons while the coal-fired thermal power units received a renovation for ultra-low emission of 700 million kilowatts, accounting for 71% of the total; more than 200,000 small coal-fired boilers were eliminated, together with over 20 million heavy-emission and old cars. In Beijing, Tianjin and the twenty-six cities mentioned above, the transformation of coal to gas and coal to electricity was completed in more than 4.7 million families, playing a positive role in improving air quality.

However, we should realize the fact that China still relies on wind to blow away smog, and the way ahead is long for air pollution control. An alarm was enough to awaken us when Beijing was hit by rounds of heavy pollution before and after the sessions of NPC and CPPCC between March 9 and 14, 2018, when a meteorological condition of temperature inversion appeared, which was the strongest one in twenty years with long duration and wide influence, making it difficult for pollutants to diffuse. Now, the concentration of PM2.5 has been lowered, significantly thanks to the targeted measures taken. Currently, China has waged a three-year campaign, which is known as the "war to defend the blue sky," to address areas with relatively heavy pollution, such as the Beijing-Tianjin-Hebei region and its surrounding areas, Yangtze River Delta, Fenhe-Weihe River Plain and so on. We concentrate on solving the problems of imbalanced industrial, energy and transportation structures through joint efforts with focus on the removal of weather with heavy pollution. This war is a tough and protracted one. Only with greater efforts and perseverance can we achieve the goals of blue sky, green earth and clear water.

The Central Economic Working Conference held in December 2017 and the *2018 Report on the Work of the Government* delivered by Premier Li Keqiang demanded that we resolutely solve the most striking problems in order to achieve greater results in the prevention and control of pollution and build a beautiful China with blue sky, green earth and clear water. It is the solemn commitment of the CPC to construct a moderately prosperous society in every respect by 2020, to which the environment quality is the key. Since the Eighteenth CPC National Congress, the Central Committee has formulated and promulgated three action plans in order to combat pollution, with following practice fully proving that the direction, path, measures, objectives and tasks of the measures in the three action plans are scientific and rational with strong pertinence and effectiveness. The Nineteenth CPC National Congress declared the "pollution prevention and control" as one of the three major battles to win in building a moderately prosperous society in an all-round way. We should also realize that it is impossible for our country to restore environmental integrity overnight, which could only be achieved with greater effort.

4.2. The experience accumulated by developed countries in handling environment and development are worth drawing on

On the whole, the environmental protection of our country is still repeating the old practice of "pollution going before abatement," with the prevention and control efforts in key areas and cities to begin only after the pollution is serious to a certain extent. Environmental pollution is due largely to the development stage of China. Except for factors such as the high proportion of heavy industries, uneven technical capacity and poor staff awareness, the long-term mechanism of policies, regulations, standards and public participation has not been formed yet. Currently, the primary energy is still the coal, the development and utilization of which will not only damage environment but also emit a large amount of sulfur dioxide, dust and other pollutants, as well as greenhouse gases such as carbon dioxide and so on. Besides, some local governments have strong economic ambitions but are slow in changing the extensive development mode while some policymakers are not fully aware of the complexity of environmental pollution and its treatment.

Learning from the experience of developed countries in dealing with environmental and development issues can help us avoid detours and save money in governance. There are many articles summarizing the practices and experience of environmental protection abroad from the perspective of legislation, policy, technology, governance mode and management. Now on the whole, the developed countries have gradually deepened their understanding of environmental protection, with the solution to pollution problems largely depending on the system design in general while the treatment of some pollutants on technological advancement. No identical pattern is universally implemented. Generally speaking, the developed countries have experienced the following five stages in their understanding of development and environmental issues.

In the first stage, environmental pollution was looked upon as the negative impact of economic growth. Polluters saw environmental protection as an unnecessary measure taken by the authorities who made storms out of a teacup. As a result, enterprises were resistant to environmental governance. Measures to control pollution discharge at this stage were "end-treatment," and were considered by the majority to be an increase in production costs.

In the second stage, people viewed environmental pollution as an integral part of production costs. At this point, polluters were beginning to recognize the potential benefits of reducing pollution intensity and took steps to reduce waste emissions during production.

In the third stage, environment was considered as one of the factors for decision-making. Polluters must take environmental factors into account when making new investment plans, and passively adopted different measures to protect the environment in the production and consumption of products.

In the fourth stage, the environment was regarded as an extremely important factor. When polluters improved their economic activities, different ecological or environmental designs were used, and in other words, systematic measures were taken to protect the environment.

In the fifth stage, the environment became one of the goals of development. People took the solution to environmental issues as the goal of social and economic policies, which guide the changes in production and consumption patterns, and structural measures were also executed to change people's attitudes toward the environment.

After the UNCED in 1992, some countries introduced economic measures which mainly included environmental taxes, emissions trading, the abolition or application of subsidy, mortgage and so on, while some others pursued compensation, liability insurance and other means to raise funds for environmental protection. No matter whether they imposed system arrangement or technical solutions, the developed countries all proceeded from the development stages, personnel quality, technical competence and so on, which requires us to combine the experience of foreign counterparts with our national conditions.

In general, China is a big developing country, whose historical task of industrialization and urbanization is far from being completed. Objectively speaking, economic development inevitably consumes a large amount of resources and energy and discharges a huge quantity of wastes, because it is impossible to build high-rise buildings or infrastructure without using resource and discharging wastes, as the technology now or even after the completion of industrialization still cannot fully convert resources into useful products. This is also why environmental protection is a long-term task.

4.3. Environmental pollution should be addressed by making more use of market mechanism

In essence, using market mechanism to solve environmental pollution is to balance the interests of all parties concerned and prevent the external diseconomy of enterprise activities, such as gaining extra profit by passing on the cost of pollution control to others. By using funds more efficiently, we can enhance environmental quality with less investment. Besides, "whoever pollutes pays" is not only a basic principle in pollution control internationally but can also give a better play to market and make pollution control more specialized with a larger scale, which is a critical way to improve efficiency.

Pollution control cannot be done at all costs. Some power plants in China adopted an "expensive" technical route for desulfurization but ended up in the inability to operate due to the high cost. Pollution control cannot be done without costs either. Some desulfurization facilities failed to stand the test of operation, due largely to the policy that "the bidder with the lowest price wins" and the competition among enterprises who tried to offer lower price by reducing necessary costs. In contrast, the total social expenditure can be saved if we use low-sulfur coal or the washed coal in which the impurities have been removed with the front-end measures. Another way to improve environmental quality is to use standards, for which an effective demonstration is Beijing, a city that has gradually set higher vehicle emission standards since 2008, relieving people of the worry that higher standards might not be able to generate the results desired. Now, the APEC Blue (the good weather during the APEC Summit in Beijing in 2014) has raised local residents' expectation of blue sky.

There is plenty we can do, such as improving the pollutant emission permit system and promoting a trading system for the amount of energy saving, pollution discharge, carbon emissions and water rights, and attracting more funds from the capital market into the building of ecological civilization. Besides, we can give full play to public financial funds as the "seed" to channel social capital to ecological conservation and environmental protection. Specifically speaking, the pollutant discharge permit system is of environmental management in nature, which allocates in writing the legitimate and specific amount of pollutants to be discharged by an enterprise according to law and is the basis for law-abiding, law enforcement and social supervision. The implementation of such permit system is conducive to putting into place the environmental protection laws and regulations, total emission reduction responsibilities and environmental protection technical specifications, and thus it is also helpful to environmental protection agencies' supervision according to law. A permit system covering major pollutant discharge, which is uniform and fair in the whole country, should be established in order to create a foundation and favorable conditions for the development of the emission trading market. It would be difficult to realize the virtuous circle of ecological system if enough attention is not paid to the improvement of people's livelihood in the process of building ecological civilization with the urgent environmental pollution problems that affect residents' lives remaining unsolved. Therefore, economic development and environmental protection shall proceed hand in hand, which means we should safeguard environmental integrity without forgetting about raising people's living standards. Only in this way can we embark on a road of civilized development featuring efficient production, prosperous life and good ecology.

4.4. The technical route should be clarified to improve the modernization of environmental governance

The main ways to reduce pollutant emission are through system, technology, engineering or management. Therefore, we should promote a green industrial development by optimizing and upgrading industrial structures and accelerating the development of advanced manufacturing industries, high technologies and modern service sectors. At the same time, we should practice the comprehensive utilization of energy, water, land, materials and resources with strict implementation of industrial policies and environmental laws and regulations, and in the meanwhile phase out production capacity of high-consumption, high-emission and low-efficiency, and prohibit the construction of new projects that waste resources and pollute environment. We should reduce the adverse impact of energy production and consumption on the environment by enhancing energy structure, controlling the total amount of coal consumption and vigorously tapping new and renewable energies. In addition, we should strengthen the development of environmental service industry, where waste generation and discharge can be reduced from the very source either through third-party treatment and replacing treatment with utilization. We should also treat wastewater with the theory of "mutual promotion and restraint," which means that we can adopt physical, chemical and biological means on one hand, and economically applicable wetland and oxidation pond and so forth on the other hand. Last but not least, we can explore the PPP mechanism. The National Development and Reform Commission and the Ministry of Finance issued circulars, respectively, demanding to establish PPP library and advance PPP projects, so as to form the industrial structures, production modes and consumption models conducive to resource conservation and environmental protection.

But we should watch closely the economic rationality of environmental technologies while promoting technical progress and selecting scientific routes. For example, problems arose when China was changing coal to gas and electricity in 2017. Between December 15 to 20 of the same year, the Ministry of Environmental Protection organized 2,367 persons in 839 groups to comb the 25,220 villages under the jurisdiction of 2,590 towns and streets of 385 counties (cities and districts) in twenty-eight cities. They found that 4.74 million out of 5.53 million households in the 25,220 villages had completed changing from coal to gas and electricity, but the problem of insufficient heating did exist. In order to protect people from the cold winter, the Ministry of Environmental Protection organized a campaign of "large-scale visit, intensive help and strict supervision," urging local governments and gas companies to solve the problem immediately after receiving reports either from inspectors who were stationed at the villages or from the villagers through telephone, WeChat or online complaints. Despite all these efforts, negative social voice was deafening. On the whole, natural gas is a scarce resource in China. It is especially true when the inelastic demand for natural gas is increasing constantly. Admittedly, for a relatively long time in the future, the promotion of "coal to gas" may produce better environmental benefits, but it is also necessary to take into account the ability to supply energy and resources as well as the economic affordability of the public. In other words, environmental protection should be considered in the overall framework of economic and social development in order to achieve a virtuous cycle between socioeconomic aspects and resources or environment.

We shall promote the organic linking between environmental protection industry and circular economy. For instance, desulfurization falls into the category of environmental protection industry, while the utilization of desulfurized gypsum is circular economy. Similarly, water processing belongs to environmental protection industry, but the use of silt from water treatment plant is a typical example of recycling economy. Therefore, the organic linking between pollutant treatment and the comprehensive exploitation of resources is key to the betterment of environmental quality. That is why we should look beyond what we see by beginning with ecologic design and clean production to achieve the transformation from the "end treatment" to process control and source reduction, from relying on administrative measures to the organic combination of legal, market and necessary administrative measures, and thus enter the new era of ecological conservation by developing a new pattern of industrialization featuring lower environmental costs, better economic returns, higher quality and sustainable growth.

4.5. Efforts should be made to encourage public participation and action to form a social atmosphere featuring resource conservation and environmental protection

It is difficult and even impossible to win the fight against pollution if the development mode and lifestyle are not green and sustainable. So, we must strengthen the protection and restoration of ecosystem to foster its service function, which is critical to the building of ecological conservation. Moreover, we should speed up the modernization of environmental governance system and capacity. For greater and better results, we should release more momentum and dividends through reforms to back the fight against pollution.

Environmental protection requires public participation. The *Opinions on Further Promoting the Development of Ecological Civilization* promulgated by the CPC Central Committee and the State Council proposed to encourage active public participation within a better public participation system, timely and accurately disclose various environmental information within a wider scope of publicity so as to defend the public's right to know and their rights to healthy environment.

The *Measures for Public Participation in Environmental Protection* came into force as of September 1, 2015. According to Article eleven, citizens, legal persons and other organizations that discover any action by any entity or individual that causes environmental pollution or ecological damage may report through letters, faxes, e-mail, 12369 hot-line for environmental protection, government websites or any other means to the competent department. The Ministry of Environmental Protection has made public the investigation into typical cases, which enhances the public's confidence in the enforcement of environmental protection laws and regulations, encourages public involvement in addressing environmental pollution and ecological damage, and helps to further consolidate the foundation for public's fight against environmental crimes.

Environmental protection needs the public to take actions. Such improper acts as spitting and littering in public by an increasing number of Chinese tourists in foreign countries have stigmatized the public image of the Chinese nation. Hence, in addition to giving full play to the roles of media and strengthening education on environmental protection and sustainable development, we should also build our capacity to safeguard our environment and give full play to environmental protection NGOs. Besides, the general public should not only realize the close ties between ecological environment and their own lives but also take practical actions, such as taking good care of public surroundings, sorting and recycling garbage, using no "disposable" chopsticks or ultra-thin plastic bags but replacing them with recycling bags and packaging. The public should also play a supervisory role and be encouraged to report environmental violations.

5. CONCLUSION

Environmental protection is like a "protracted war" that runs through the whole process of modernization of China. To firmly establish the concept of ecological civilization in the whole society, we must make great efforts in analyzing the problems, defining strategies, assessing financial strength, strictly enforcing the law and putting the implementation in place. Only when we all work together will our living environment be better day by day and our common and only home be beautiful.

Notes

1 Editorial Committee of China Environmental Protection Administration for 20 Years. (1994). *China Environmental Protection Administration for 20 Years*. Beijing: China Environmental Science Press, p. 3.

2 Ma Hong. (Ed.). (1982). *Modern Chinese Economics Encyclopedia*. Beijing: China Social Sciences Press, p. 571.
3 Editorial Committee of China Environmental Protection Administration for 20 Years. (1994). *China Environmental Protection Administration for 20 Years*. Beijing: China Environmental Science Press, p. 3.
4 Yu Yong and Li Yan. "Qu Geping: Premier Zhou Enlai forced out the first generation of new China environmentalists", http://www.ce.cn/, accessed: April 10, 2018.
5 Ma Hong (Ed.). (1982). *Modern Chinese Economics Encyclopedia*. Beijing: China Social Sciences Press, p. 571.
6 Editorial Committee of China Environmental Protection Administration for 20 Years. (1994). *China Environmental Protection Administration for 20 Years*. Beijing: China Environmental Science Press, p. 3.
7 Zhang Kunmin (2004). *Policies and Actions on China's Sustainable Development*. Beijing: China Environmental Science Press, p. 15.
8 Sun Fangming. (2000). *A Concise Course on Environmental Education*. Beijing: China Environmental Science Press.
9 Wang Jinnan, et al. (2004). *Energy and Environment: China 2020*. Beijing: China Environmental Science Press.
10 Jiang Weixin and others (2006). *A Collection of China's Policies on Building an Economical Society*. Beijing: China Development Press.
11 China Environmental Protection Foundation (1998). *A Preliminary Study of Public Environmental Awareness in China*, China Environment Publishing Group, pp. 68–88.
12 China Environmental Protection Foundation (1998). *A Preliminary Study of Public Environmental Awareness in China*, China Environment Publishing Group, pp. 68–88.
13 Jiang Weixin and others (2006). *A Collection of China's Policies on Building an Economical Society*. Beijing: China Development Press.
14 Jiang Weixin and others (2006). *A Collection of China's Policies on Building an Economical Society*. Beijing: China Development Press.
15 "The State Environmental Protection Administration sets up 11 supervising authorities for law enforcement", page on environmental protection in people.cn, July 31, 2006.
16 Editorial Committee of China Environmental Protection Administration for 20 Years. (1994). *China Environmental Protection Administration for 20 Years*. Beijing: China Environmental Science Press, p. 3.
17 China Trans-century Environmental Protection Inspection Campaign Executive Committee, *China Trans-century Environmental Protection Inspection Campaign Activities (1993–2007)*, an internal material.
18 The Current Status of NGO for Environmental Protection in China is Worrying: Negotiation Mainly Depends on the Government. www.chinagate.com.cn, accessed: April 25, 2006.
19 Yu Yong and Li Yan. "Qu Geping: Premier Zhou Enlai forced out the first generation of new China environmentalists", http://www.ce.cn/, accessed: April 10, 2018.
20 Li Ganjie. "Do a Good Job in the Uphill Battle for Prevention and Control of Pollution", http://news.cctv.com/2018/03/17/VIDE0QN8S7ESBLoH35vKqH12180317.shtml, accessed: March 17, 2018.
21 Sun Youhai (2008). The Basic Experience and Existing Problems of Environmental Legislation in China since the Reform and Opening Up, *Journal of China University of Geosciences (Social Science edition)*, No. 4.

22 Li Ganjie. "Do a Good Job in the Uphill Battle for Prevention and Control of Pollution", http://news.cctv.com/2018/03/17/VIDE0QN8S7ESBLoH35vKqH12180317.shtml, accessed: March 17, 2018.
23 Yu Yong, Li Yan, "Thanks to Premier Zhou Enlai, the first batch of environmental protectors emerged in China: Qu Geping". China Economic Net, accessed: April 10, 2018.
24 Dai Zhengzong, "Central government supports the building of a beautiful China". China Financial and Economic News, March 23, 2018.
25 Li Ganjie. "Do a Good Job in the Uphill Battle for Prevention and Control of Pollution", http://news.cctv.com/2018/03/17/VIDE0QN8S7ESBLoH35vKqH12180317.shtml, accessed: March 17, 2018.

CHAPTER THREE

Environmental Protection Planning: Review and Outlook

WU SHUNZE, WAN JUN, YANG LIYAN AND ZHAO ZIJUN[*]

1. INTRODUCTION

Since the adoption of the reform and opening-up policy in 1978, China has registered rapid economic and social development. With the deepening understanding of ecological and environmental issues, the country has also advanced its environmental planning alongside. In 1976, China released its first ten-year plan for environmental protection. During the Seventh *Five-Year Plan* period (1986–1990),

---

[*] Wu Shunze, doctoral supervisor, research fellow, Director and Party Secretary of the Policy Research Center for Environment and Economy of the MEE, Deputy Director of the Reform Office of the MEE, Deputy Director of the Vertical Reform Office of the MEE, whose long-time research interests include national-regional / basin-urban environmental protection strategy, planning, engineering, policy and so on. Wan Jun, Ph.D., research fellow, Director of the Environmental Strategy Institute of the Chinese Academy of Environmental Planning of the MEE, member and Secretary-General of the Special Committee on Environmental Planning, whose research interests include eco-environment planning, urban environment and ecological evaluation. Yang Liyan, a graduate student and assistant researcher of the Chinese Academy of Environmental Planning of the MEE, whose research interests include environmental protection planning, environmental science and resource utilization and so on. Zhao Zijun, a graduate student and intern of the Policy Research Center for Environmental and Economy of the MEE, whose research areas include green development, eco-environmental protection policy and international environmental cooperation.

China's environmental efforts were largely centered around the control of industrial pollution. Then during the Tenth *Five-Year Plan* period (2001–2005), environmental protection planning gained momentum on a gradual basis when the nation began to lay focus on restoring environmental quality. The Eighteenth CPC National Congress stipulated that the building of ecological conservation be incorporated into the "Five-sphere Integrated Plan," which is the overall layout for the building of socialism with Chinese characteristics, paving a path for the further development of the country's environmental protection planning. Since the Nineteenth CPC National Congress, the CPC Central Committee with comrade Xi Jinping at its core has raised ecological conservation and environment protection to an unprecedented height, leading to dramatic changes in the nation's environmental protection cause. At present, the coverage of environmental protection planning in China has been gradually expanded to multiple aspects of ecological environmental protection.

This chapter presents a review of the economic, social and environmental backgrounds and development course of China's environmental protection planning in the four decades of reform and opening-up, an analysis on the determinant factors of environmental protection planning, and a prospect on its future development.

2. THE ECONOMIC, SOCIAL AND ENVIRONMENTAL BACKGROUNDS OF THE DEVELOPMENT OF ENVIRONMENTAL PROTECTION PLANNING

China's reform and opening-up in its constant expansion has laid emphasis on the modification of the institution, mechanism and policy in the economic arena. Along with the rapid progress of society and economy, environmental complications were also gaining prominence. The relationship between development and ecological protection has been adjusted on a frequent basis in a trend of spiral evolution.

2.1. Economic and social development since the reform and opening-up

Earth-shaking changes have taken place in the country's economy and society. In the forty years from 1978 to the end of 2017, the Chinese economy grew by 226 times with per capita GDP increasing by 186 times, the urbanization rate rising from 17.8% to 58.5%, and the economic aggregate growing from the tenth place in the world to the second (see Figure 3-1). As reform and opening-up moves

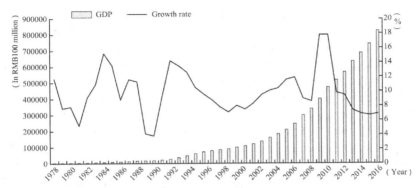

Figure 3-1 China's Economic Growth Rate since the Reform and Opening-up

further and deeper, the national governance system and capacity have also undergone tremendous changes.

In the forty years when China opened wider to the world, there had been four key time nodes in its economic and social development. The first is the Third Plenary Session of the Eleventh Central Committee of the CPC held in 1978, which marked China's comprehensive launch of reform and opening-up, with the country beginning to "cross the river by feeling the stones" and exploring a socialist road with Chinese characteristics. The second is 1992 when the Fourteenth CPC National Congress established the reform goal of building a socialist market economy, in which the market would play a critical role in the allocation of resources. The third is the final signing of China's accession to the WTO in 2001, when Chinese economy began to integrate with the world in an all-round way amid global division of labor. The fourth is 2012, the year that witnessed the convention of the Eighteenth CPC National Congress, which stipulated that ecological conservation should be incorporated into overall layout of the "five-sphere integrated plan" and the strategic layout of "four-pronged comprehensive strategy."[1] Currently, China's economy has entered a new phase featuring a transformation from high-speed growth to high-quality development.

2.2. Reform and development of the eco-environment protection system

During the rapid development of economy and society, the environmental problems of China demonstrated the characteristics of being compound, compressive and accumulative, due largely to the fact that it took China only four decades to finish the urbanization, industrialization and globalization that had taken western countries more than a century. Accordingly, the environmental problems that once plagued the developed nations for over 100 years came to China all at the same time, a situation which was further complicated by the fact that the country is not only large in areas but also has a wide gap in development levels in different regions. Thankfully, China's reform and opening-up has not only breathed vitality into economy but also facilitated the country in adapting its systems, mechanisms and policies to the needs of economic and social development. As an integral part of national governance system, environmental governance has also undergone enormous changes in the past forty years. China has held seven national environmental protection conferences according to the stages the country is currently in, which, on top of a great array of major systems and policies introduced, represents the CPC's and the country's strategic positioning and deployment of ecological environment.

The Chinese delegation participated in the first UN Conference on the Human Environment held in Stockholm, Sweden, on June 5, 1972, which made our government begin to realize the importance of environmental issues.[2] Between August 5 and 20, 1973, the State Development Planning Commission organized on behalf of the State Council the first National Environmental Protection Conference in Beijing, which adopted the *Several Provisions on Protection and Improvement of the Environment*, and laid down the "Thirty-two-Character Principles" of "overall planning and rational distribution; utilizing resources comprehensively and turning hazards into benefits; relying on the people and involving everybody; and protecting environments for the benefit of the people." This conference successfully called the attention of government officials at all levels to the problem of environmental protection.

In December 1983, the convention of the second National Environmental Protection Conference was organized, which summarized the achievements and experiences of environmental protection in the past ten years and studied the principles, policies and objectives to be implemented in the future. Whereas China's population and resources were not in harmony with each other, the conference adopted environmental protection as a basic state policy. The principle of "three synchronizations and three unifications" proposed at the conference shows that China began to recognize the innate relationship between environment, economic development and urbanization, producing a great and far-reaching influence on the development of environmental planning.[3]

The sustained development of national economy and society renders an ever higher strategic position to environmental protection, which was reflected in the following national environmental protection conferences.[4] See Table 3-1 for details.

The National Conference on Ecological and Environmental Protection was held in Beijing between May 18 and 19, 2018. The conference established Xi Jinping Thought on Ecological Civilization, stressing that we must strengthen

Table 3-1 The Positioning by the National Environmental Protection Conferences (NEPC) of the Role of Environmental Protection in Economic and Social Development

NEPC	Time	Landmark Achievements	Relevant Expressions
First	Aug. 1973	For the first time, China has found the environmental problems in the country, which caught the attention from all walks of life, and formulated environmental planning.	The *Several Provisions on Protection and Improvement of the Environment* was adopted, laying down the "Thirty-two-Character Principles" of "overall planning and rational distribution; utilizing resources comprehensively and turning hazards into benefits; relying on the people and involving everybody; and protecting environments for the benefit of the people." This is China's first strategic policy on environmental protection.
Second	Dec. 1983	Environmental protection has become a basic state policy	The policy of "three synchronizations and three unifications" was adopted, on top of the three major policies of "prevention first; prevention and control integrated," "whoever pollutes is accountable for remediation" and "strengthening environmental management." It was proposed that environmental protection should be included in the national economic and social development plan.
Third	Apr. 1989	The eight environmental management systems have been established	The eight environmental management systems with Chinese characteristics were defined, including "environmental impact assessment," "synchronous planning, implementation and development," "charge of pollution discharge," "responsibility for environmental protection objectives," "quantitative assessment of comprehensive urban environmental remediation," "registration and permit for pollution discharge application," "remediation within time limit" and "centralized control."

(continued)

Table 3-1 Continued

NEPC	Time	Landmark Achievements	Relevant Expressions
Fourth	Jul. 1996	The idea that environmental protection is key to the implementation of the sustainable development strategy, and protecting environment is protecting the productive forces have been adopted	The State Council made the *Decision on Several Issues Concerning Environmental Protection*, specifying the objectives, tasks and measures of trans-century environmental protection. The policy of paying equal attention to pollution control and ecological conservation was determined, with large-scale pollution control and environmental protection projects in key cities, river basins, regions and sea areas beginning to be carried out throughout the country. Environmental protection entered a new stage.
Fifth	Jan. 2002	The idea that environmental protection is an important function of government has been advocated	It was pointed out that environmental protection is an important function of the government and the whole society should be mobilized in accordance with the requirements of the socialist market economy. The theme of the conference was to implement the *Tenth Five-Year Plan for National Environmental Protection* approved by the State Council and to deploy the work related to environmental protection during the Tenth Five-Year Plan period.
Sixth	Apr. 2006	The Scientific Outlook on Development has been implemented and environment protection has been put in a more important strategic position, so as to realize the three historical changes.	Environmental protection is in the interest of the overall and long-term development of China's modernization; thus, it is a cause that benefits all generations. We shall put environmental protection in a more important strategic position, and do it well in the spirit of being highly responsible to the country, to the nation and to the future generations.

Table 3-1 Continued

NEPC	Time	Landmark Achievements	Relevant Expressions
Seventh	Dec. 2011	The idea of protection in development and development in protection was adopted	In accordance with the requirements of the Twelfth Five-Year Plan, we will adhere to the principle of "protection in development and development in protection," promote economic transformation, improve the standard of life, consolidate the foundation for long-term stable and rapid economic development, and provide people with livable and healthy environment.

the leadership of the Party in the building of ecological civilization. The meeting proposed that efforts be stepped up to push the development of ecological civilization to a new height by addressing ecological damages and waging a tough fight against pollution.

2.3. The evolution of environmental planning in the process of reform and opening-up

Since China carried out reform and opening-up and shifted its focus to economic growth, it has undergone a fundamental transformation in its economic system from a highly centralized planned economy to a socialist market economy, attaining remarkable achievements. What is worth mentioning is that Chinese economy grew by more than 10% annually for many consecutive years.[5] However, problems like forest reduction, soil erosion and desertification emerged along with the growth of population and economy.

Environmental protection planning is the ruling system for environmental protection. Its focus is constantly adjusted in line with the prominent environment issues and the public understanding of the ecological conservation at that time. For example, during the Seventh and Eighth Five-Year Plan periods, the central work of environmental protection planning was the treatment of industrial pollution.

During the Ninth Five-Year Plan period, China decided to pursue sustainable development, which gained momentum in the environmental planning of the Tenth Five-Year Plan. Then, during the Eleventh Five-Year Plan period, the government put environmental protection in a more important strategic position

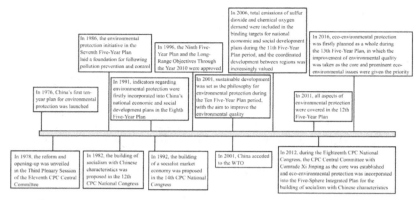

Figure 3-2 Roadmap of China's Environmental Planning and Socioeconomic Development Stages

by taking the significant reduction in the total emission of major pollutants as a binding indicator for economic and social development, with focuses laid on addressing outstanding environmental issues. The Eighteenth CPC National Congress incorporated the building of ecological conservation into the "five-sphere integrated plan" for building socialism with Chinese characteristics. President Xi Jinping established the theoretical basis of ecological civilization with his philosophy that "lucid waters and lush mountains are invaluable assets," reflecting his concept of governing the country through sustainable development.[6] These national policies and state-governing concepts are embodied in the environmental planning during the Twelfth Five-Year Plan and the Thirteenth Five-Year Plan (see Figure 3-2 for details). It is safe to say that the development of environmental planning in China is also a process where the country has pursued environmental integrity from scratch to establishing a mature and complicated system, from solving local accidents to development in an all-round way, with different emphases in different development stages.[7]

3. THE FIRST STAGE: 1978–1992

The environmental protection planning in China entered its initial stage of development from 1978 when the reform and opening-up policy was adopted to 1992 when the Fourteenth CPC National Congress proposed the reform goal of building a socialist market economic system, which marked the beginning of the country's embarking on the path of socialism with Chinese characteristics.

3.1. The starting stage (1978–1982)

In 1978, the Third Plenary Session of the Eleventh CPC Central Committee explicitly shifted the focus of the Party's work to the socialist modernization drive, and announced that China would begin to implement the policy of internal reforms and opening to the outside world, which symbolized the new period in Chinese history featuring the country's integration into the world.

During this period, however, ecological integrity was not a priority. The first National Environmental Protection Conference awoke public awareness of the domestic environmental situation and problems, signifying the start of environment management in the country. In 1978, when it endorsed the work report delivered by the Environmental Protection Leading Group under the State Council, the CPC Central Committee pointed out that it was an integral part of socialist construction and modernization drive to eliminate pollution and protect environment, and China would never take the old wrong road of treatment after pollution. This is the first time in the history of the Communist Party of China to give important instructions on environmental protection in the name of its Central Committee.

But China's environmental efforts was at its early stage, as environmental surveys and assessment were only carried out in some areas, and the environmental protection planning on a deeper and large-scale basis had not been started yet. The exploration into environmental planning at that time was mainly to deepen public understanding of environmental protection.

During the same period of time, the National Economic Development Plan (which was often referred to as the Fifth Five-Year Plan for 1976–1980) proposed that large- and medium-sized industrial and mining enterprises as well as the enterprises with serious pollution hazards strictly treat the "three wastes" (waste water, gas and residue) according to the discharge standards set by the state. In addition, industrial and domestic sewage was processed in accordance with the national regulations in the eighteen key cities of environmental protection priority including Beijing, Shanghai and Tianjin. Major ports and river systems such as the Yellow River, Huaihe River, the Songhua River, the Lijiang River, Baiyang Lake, Guanting Reservoir and the Bohai Sea witnessed better pollution control and improved water quality.[8]

Environmental protection was then incorporated into the Sixth Five-Year Plan as an independent chapter (Chapter Thirty-five: Environmental protection) and ranked among the ten basic tasks along with the development of industrial and agricultural production, adjustment of product structure, and encouraging technological transformation and expansion of foreign trade, giving rise to the proposal of establishing the "three synchronizations system" (the synchronous planning, implementation and development of economy, urban and rural construction and environmental protection) and improving the "three wastes" treatment capacity. At the same time, many regions and cities began to draw up environmental planning based on the methods introduced by foreign countries. The representative environmental planning included "Shanxi Environmental Planning for Energy and Heavy Chemical Bases," "Ji'nan City Environmental Planning," "Changchun City Environment Planning" and so forth.[9]

Every city did much work in accordance with the environmental objectives set by the Sixth Five-Year Plan and made great achievements. For example, the implementation rate of the "environmental impact statement" system and the "three synchronizations system" reached 90% and 85%, respectively. The rapid deterioration of the urban environment was contained, with environmental management strengthened.

3.2. The stage featuring all-round development of economy (1983–1992)

Years of practice have proven that planned economy alone could constrain the development of productive forces. Since China implemented the reform and opening-up policy, one pressing issue confronting the country was the relationship between a planned economy and a market economy and that between socialism and capitalism. With the deepening of reform, Deng Xiaoping proposed to "develop a market economy under socialism," which provoked wide thoughts. In 1982, the Twelfth CPC National Congress put forward the proposition of "building socialism with Chinese characteristics," which then became a basic thought guiding China's socialist drive in the period of reform and opening-up and unshackled the thinking that planned economy was the opposite of market regulation. Besides, it put forward the principle to maintain the planned economy as the mainstay of Chinese economic system and market forces as a supplementation. Since then, economic system reform was carried out in an all-round way in China.[10]

In 1992, the Fourteenth CPC National Congress explicitly put forward the goal of reform to establish a socialist market economic system and that market should play a fundamental role in resource allocation under the state's macro-control. Thanks to the rapid economic and social development, people were able to see the hope of being lifted out of poverty and take their first bite into the benefits brought by reform and opening-up. However, in the transition period from planned economy to market economy, the traditional extensive growth mode had been dominant with a strong inertia. Urban environmental pollution has become the center of environmental problems in China.

In accordance with the *Proposal of the CPC Central Committee on Formulating the Seventh Five-Year Plan for National Economic and Social Development*, the Environmental Protection Committee under the State Council organized and prepared the first draft of the *National Environmental Protection Plan* for the Seventh Five-Year Plan period, which, after much calculation, balancing, demonstration, research and revision, was reviewed and adopted in principle at the ninth meeting of the Environmental Protection Committee under the State Council. Thus, the five-year plans for environmental protection have become an integral part of China's national economic and social development plans.

The national environmental protection plan highlighted the comprehensive remediation of urban environment and the prevention and control of industrial pollution, which, on top of requiring that different regions and industries should put forward their own pertinent environmental protection objectives, laid emphases on the constraint and total amount control of environmental capacity by demanding that industries in areas with high population density and industrial concentration be gradually relocated to areas with larger environmental capacity. According to the plan, natural resources could only be exploited when the conditions of environmental capacity permit while at the same time, research on environmental capacity should be continued. The plan also requested to gradually control the total amount of pollutants in the water system jointly used by economic zones and urban agglomerations with equal attention paid to the new environmental problems that had arisen in economic zones, urban agglomerations and township enterprises and to the major role of environmental management system in environmental protection plans. As a result, environmental pollution was abated to a certain extent against the backdrop of insufficient environmental remedy funds from the state on one hand and the boisterous economic expansion on the other hand.

The environmental protection plan for the Seventh Five-Year Plan period was significantly more scientific than that for the Sixth Five-Year Plan period. For example, it was the first of its kind to possess such rich contents, complete indicators and systematic methods. During this period, the government began to link environmental planning with environmental management by trying out in some cities the pollutant discharge permit system, where such permits were issued for better control of pollution.

The *Ten-Year Plan for Environmental Protection and Outline of the Eighth Five-Year Plan* (hereinafter referred to as the *Outline of the Eighth Five-Year Plan*) was based on the *Proposal of the CPC Central Committee on Formulating the Ten-Year Plan for National Economic and Social Development and the Eighth Five-Year Plan* as well as the *Ten-Year Plan for National Economic and Social Development and Outline of the Eighth Five-Year Plan of the People's Republic of China* and was prepared by relevant departments of the State Council and different regions organized by the National Environmental Protection Agency under the guidance of the State Development Planning Commission.

The *Outline of the Eighth Five-Year Plan* strengthened the requirements of the Seventh Five-Year Plan for the control of total amount, proposing that the focus of pollution control be gradually shifted from concentration control to total amount control, from end control to that of whole process, with equal attention paid to the control of the total amount of industrial dust discharge and that of total pollutant discharge in key industrial pollution sources, river basins and sea areas. The *Outline* called for attention to the coordinated development of environmental protection, economy and society, as well as environmental management and sci-tech progress. During this period, remarkable results have been achieved in the prevention and control of industrial pollution and the comprehensive urban environmental control, as well as in the construction of natural reserves. Thus, environmental protection has been incorporated into national economic and social development planning at all levels along with the preliminary formation of an environmental planning system which integrates the planning of macro environmental protection targets with the planning of environmental quality, with the former aiming to promote the sustainable and coordinated development of economy and environment and the latter featuring the control and treatment of pollutant discharge at the source. This is an important symbol of environmental planning entering the development stage.

4. THE SECOND STAGE: 1993-2000

China's industrial reform and economic transformation went faster and deeper in the 1990s, when its GDP grew by 10.4% annually, signifying the accelerating industrial sectors and the all-round development of socialist market economy in the country.

In the 1990s, while strengthening the prevention and control of pollution caused by enterprises, China also beefed up on a large scale the abatement of non-point source pollution in rural areas and the environmental governance in key cities, river basins and regions. However, along with the sustained and rapid development of Chinese economy, the environment woes that once plagued Western countries in stages in their century-long industrialization hit China all at once, leading to increasingly prominent contradictions between environment and development, represented by the rising discharge of organic pollutants.

To implement the Ninth Five-Year Plan for National Economic and Social Development and Outlines of Objectives in Perspective of the Year 2010 and the Decision of the State Council on Several Issues Concerning Environmental Protection (hereinafter referred to as the Decision), the State Council convened the Fourth National Environmental Protection Conference in July 1996, deliberated and adopted the National Ninth Five-Year Plan for Environmental Protection and Outlines of Objectives in Perspective of the Year 2010 (hereinafter referred to as the Ninth Five-Year Environmental Plan), the first of its kind to be approved by the State Council.

The UN Conference on Environment and Development held in 1992 prompted *China's Ten Countermeasures for Environment and Development*, which requested that people's governments at all levels and relevant departments prepare environmental protection plans when formulating and implementing the development strategy and that sustainable development be pursued. The *Ninth Five-Year Environmental Plan* further clarified the strategy of sustainable development and listed the environmental protection goal of sustainable development in the national economic and social development plan.

On top of its effort to basically control the trend of exacerbating environmental pollution and ecological damage, the *Ninth Five-Year Environmental Plan* did improve environmental quality of some cities and regions, with its proposal to "set the stage for the control of total pollutant emission." The *Ninth Five-Year Environmental Plan* was an important measure for the implementation of the *Decision*, which stipulated that "by the end of year 2000, all provinces, autonomous regions and municipalities directly under the central government should control the maximum discharge of major pollutants under their respective jurisdiction within the maximum discharge quantity index laid down by the state; industrial pollution sources all over the country shall meet national or local standards on discharge of pollutants; and cities should all keep the quality of air and surface water up to the corresponding national standards in terms of their functional categories." The *Ninth Five-Year Environmental Plan* also included the treatment of three lakes (Taihu Lake, Chaohu Lake, Dianchi Lake), three rivers (Huaihe River, Haihe River, Liaohe River) and two controlled areas (area of acid rain control and area of sulfur dioxide pollution control), one city (Beijing) and one sea (the Bohai Sea). This is called the "33211 Project" in short. When China adopted the principle of advancing pollution abatement with ecological conservation, two documents, namely, the *National Plan for Total Amount Control of Major Pollutants Discharge in China during the Ninth Five-Year Plan Period* and *China Cross-century Green Projects* represented the country's major moves, which, to certain extent, constituted an innovation of and a breakthrough in previous environmental protection plans during the Seventh and Eighth Five-Year Plan periods.

During this period, environmental protection witnessed a scenario of multi-sectoral cooperation and participation. The National Environmental Protection Agency prepared six national environmental plans, which were approved by the State Council, and the *China Biodiversity Conservation Action Plan* and the *Outline of China Natural Reserve Development Plan (1996–2010)*, which were jointly released with thirteen departments including the State Development Planning Commission.

5. THE THIRD STAGE: 2001–2011

After fifteen years of negotiation, China became the 143rd official member of World Trade Organization on December 11, 2001, starting to fully engage itself in international labor division. Along with another round of boom in its economy, China began to explore how to achieve new industrialization. In the decade from 2001 to 2010, China's GDP grew at an average annual rate of 10.5%,[11] higher than the 9.3% in the 1980s and 10.4% in the 1990s. China's GDP increased from RMB12 trillion in 2002 to nearly RMB52 trillion in 2012. The breakdown of primary, secondary and tertiary industries in 2012 stood at 9.4: 45.3: 45.3. Compared with 2002, the share of the primary industry continued to drop while the share of secondary and tertiary industries rose, signifying a better industrial structure. The labor force that used to engage in the primary industry was then transferred to the modern and efficient secondary and tertiary industries. The industrial structure was in a process of gradual upgrading and transformation, with social labor productivity as a whole being effectively elevated.

The Seventeenth CPC National Congress clearly proposed to holistically understand the new situations and tasks involved in the further development of industrialization, IT technology, urbanization, marketization and international integration, rendering market a clearer and more accurate positioning. By proactively following the trend of global industrial division, China gave full play to its comparative advantages and accepted many industries transferred from other countries, which helped to boost its export-oriented economy where foreign trade and bilateral investment were momentously promoted and encouraged. As a result, China's open economy registered a huge upsurge, pushing the country's national strength to one new high after another.[12] Thanks to the upgrading of resident consumption structure, the capital and technology-intensive industries like the automobile industry also achieved rapid development and became the economic pillars of China, which further promoted the growth of steel, machinery and other relevant sectors.

During this period of time, China faced the degraded ecological system and a fragile ecological environment. Global concern over the biodiversity conservation in China also exerted increasingly huge pressure on the country when it was faced with many other problems such as the continuous expansion of population, the rising total amount of energy consumption, the accelerating industrialization and urbanization, and the increasing pollutant discharge. Chinese economy felt more constraints from environment than ever before.

The *National Tenth Five-Year Plan for Environmental Protection* (hereinafter referred to as the *Tenth Five-Year Environmental Plan*) was approved by the State Council in December 2001, requesting that environmental protection efforts be continued to address the pollution in three lakes, three rivers and two controlled areas, Beijing and the Bohai Sea, which were identified in the *Ninth Five-Year Environmental Plan* as key regions, while strengthening the treatment of water pollution in the Three Gorges Reservoir Area and along the South-to-North Water Transfer Project. The *Tenth Five-Year Environmental Plan* adhered to the basic national policy of environmental protection and followed the strategy of sustainable development, with the improvement of environmental quality as its objective to safeguard national environmental security and people's health. The *Plan* was based on river basins and administrative regions, with highlights on classified guidance.

During the Tenth Five-Year Plan period, environmental protection efforts were stepped up in all regions and by relevant stakeholders. For example, a large number of backward production capacities with high consumption and high pollution were eliminated. Pollution control and the construction of urban environmental infrastructure were sped up, along with the continuous advancement of pollution abatement in key regions, river basins and cities. Thus, all these moves strengthened ecological protection and governance.

During the Eleventh Five-Year Plan period, the State took a significant reduction in the total emission of major pollutants as a binding indicator of the outline of national economic and social development, with focus laid on protruding environmental problems and great results achieved in aspects such as understanding, policy, system, capacity building and so forth. The state established a strict system for the assessment of total emission and triggered the large-scale treatment of pollution with major projects. By 2010, chemical oxygen demand and the total emission of sulfur dioxide decreased by 12.45% and 14.29%, respectively, compared with 2005, outperforming the designated targets.

During this period of time, environmental plans of various kinds were made in an all-round way. In addition to the national plan for environmental protection and the eight special environmental plans, the State Council also issued the outline of national ecological conservation plan. The *National Plan for the Breakdown of Major Pollutant Emission Control during the Tenth Five-Year Plan Period* was compiled and implemented, defining six indicators for controlling the total amount of major pollutants and breaking down the total amount among all provinces, autonomous regions and municipalities directly under the Central Government and the cities with independent plans. On the basis of fourteen years of water environmental function zoning across the country, the National Environmental Protection Administration preliminarily completed on a national scale, for the first time, the compilation of water environmental function zoning, as well as the compilation of ecological function zoning.

Significant progress was also achieved in environmental protection planning in key areas. The *Environmental Protection Plan of the Pearl River Delta Region* and the *Comprehensive Environmental Protection Plan of Guangdong Province* were approved by the People's Congress of Guangdong Province. These two plans were known for their significant innovations in planning concepts, technologies and methods, key tasks and implementation mechanism. For example, they put forward the concept of space control for ecological environment for the first time, defining 14.13% the Pearl River Delta and 20% of Guangdong Province as ecological areas under strict control – the earliest practice of red line in ecological protection in China. Afterward, environmental protection plans for the Yangtze River Delta and the Beijing-Tianjin-Hebei region were prepared.

In 2005, the Standing Committee of the Political Bureau of the CPC Central Committee and the Executive Meeting of the State Council discussed for the first time a report on the *National Eleventh Five-Year Plan for Environmental Protection* (hereinafter referred to as the *Eleventh Five-Year Environmental Plan*), which signified the importance attached to environmental protection by the Party and country. The *Eleventh Five-Year Environmental Plan* was issued, for the first time, by the State Council. It promoted the historic transformation of environmental protection in China from the previous over-emphasis on GDP and the planning of balanced control of the total amount to the planning that focuses more on coordinated regional development, spatial layout and development quality. Viewed from the restrictions on government, its biggest characteristic is that it stresses the implementation and evaluation of planning and its rigid constraints on the government side.

When we look at how national economic planning channels the environmental protection efforts, the Ninth and Tenth Five-Year Environmental Plans as well as previous plans had their focus on specific regions or industries, which consisted of mostly the environmental protection in urban or rural areas and the prevention and control of industrial pollution. However, the *Eleventh Five-Year Environmental Plan* emphasizes the role of factors, with the management of water, air and waste residue, and so on in a classified way. National economic planning included environmental protection in several aspects, such as the promotion of coordinated regional development, in addition to independent chapters on tasks of building a resource-saving and environment-friendly society. Therefore, the *Eleventh Five-Year Environmental Plan* has more profound connotation than the *Tenth Five-Year Environmental Plan* in its statement involving population, resources and environment. It stressed the need to implement strong environmental protection measures, mainly through the improvement of laws and regulations, the strengthening of law enforcement and other legal means, supplemented by economic measures.

During the Eleventh Five-Year Plan period, China made huge efforts to solve outstanding environmental problems, with substantial progress made in public understanding, policies, system and capacity building. The environmental protection objectives and key tasks of the Eleventh Five-Year Environmental Plan were fully fulfilled. What is worth mentioning is that China has outperformed the plan in the two indicators of pollution emission reduction.[13]

In 2008, the Ministry of Environmental Protection and the Chinese Academy of Sciences jointly promulgated the *National Eco-Environmental Function Zoning* with progress made in water environmental function zoning. The *Outline of the Eleventh Five-Year Environmental Plan* explicitly stated that twenty-two important ecological functional areas be protected with top priority and be developed moderately. The State Council issued the national major function zoning in December 2010, which divided the country's land and space into four categories: areas with optimized development, areas with focus on development, areas restricting development and areas banning development. The national zoning of function areas served as a basis for the preparation and compilation of the twelfth national plan for economic and social development, as well as for other regional and urban planning.

The *National Twelfth Five-Year Plan for Environmental Protection* (hereinafter referred to as the *Twelfth Five-Year Environmental Plan*) was also printed out and distributed by the State Council. Its major indicators included, firstly, significant reduction of the total discharge of major pollutants [the chemical oxygen demand (COD) and SO2 will decrease by 8% and the ammonia nitrogen and nitrogen oxides will decrease by 10% in 2015 compared with that of 2010]; secondly, significant improvement of environmental quality (proportion of surface water sections under national monitoring program failing to meet Class V standard will be lower than 15%; and the proportion of surface water sections of the seven big water systems under national monitoring program meeting Class III standard will be greater than 60%).

Compared with previous five-year plans for environmental protection, the *Twelfth Five-Year Environmental Plan* embodies the strategic idea of "protection in development and development in protection," the historic change to pursue economic growth through environmental betterment, and the balanced and coordinated arrangement by the state for major strategic tasks in restoring ecological integrity. The guiding philosophy for environmental planning clings on scientific development and the acceleration of transformation of economic development modes, endeavoring to build a higher level of ecological conservation and to address the prominent environmental problems that undermine scientific development and public health. The *Twelfth Five-Year Environmental Plan* pushed forward the historic transformation of environmental protection in an all-round manner, proactively exploring new remedies that feature low cost, high efficiency, reduced emission and sustainable development and speeding up the building of a resource-saving and environment-friendly society. With regard to planning formulation mechanism, the government adopted an open attitude by reinforcing basic research, recruiting preliminary research institutes publicly, collecting opinions and distributing questionnaires online, carrying out planning survey and hosting brainstorming sessions nationwide, and taking suggestions from experts and scholars of different backgrounds. In the content of planning, four strategic tasks were proposed, which included, specifically, further reducing the total amount of pollutants emitted, improving environmental quality, preventing environmental risks and ensuring equal access to basic public environmental services for urban and rural areas alike. The *Outline of the Twelfth Five-Year Plan for National Social and Economic Development* incorporated all the major objectives, major indicators, key tasks, policy measures and key projects of the *Twelfth Five-Year Environmental Plan*.

During this period, in addition to the national environmental protection plan, the State Council issued three plans and approved seven more plans. To better facilitate the national major function zoning in China, the Ministry of Environmental Protection started the national environmental function zoning, which could be divided into the stage of preliminary research (2009–2010) and the stage of preparation and application (2011–2013). The Ministry of Environmental Protection chose the provinces of Hebei, Jilin, Heilongjiang, Zhejiang, Henan, Hubei, Hunan, Sichuan, Qinghai, and Guangxi Zhuang Autonomous Region, Ningxia Hui Autonomous Region, Xinjiang Uygur Autonomous Region and Xinjiang Production and Construction Corps to carry out pilot projects, and issued the *Technical Guide for National Environmental Function Zoning (Trial)*.

The Twelfth Five-Year Plan period saw the improvement in environmental quality, with the targets of pollution control and emission reduction overfulfilled. Ecological conservation and building achieved great results with environmental risks gradually being put under control and environmental legislative and governance efforts advancing constantly.[14] The pilot projects of preparing overall plans for urban environment initiated in the Twelfth Five-Year Environmental Plan period also witnessed great results. More than forty cities across the country, including Beijing, Guangzhou, Fuzhou, Chengdu, Qingdao, Ji'nan, Harbin and Urumqi, launched the preparation of their overall plans for urban environment, which were basic, strategic, spatial, harmonious and systematic in nature and had made their urban environmental governance more systematic, scientific, law-based, refined and of higher application level of information technology.

6. THE FOURTH STAGE: 2012 TO THE PRESENT

The Eighteenth National Congress of the Communist Party of China (hereinafter referred to as the Eighteenth CPC National Congress) was held on November 8, 2012, in Beijing, setting a new orientation for the construction of socialism with Chinese characteristics. Before the Sixteenth CPC National Congress, its content mainly focused on economic, political and cultural construction, and social construction was added at the Seventeenth CPC National Congress. The Eighteenth CPC National Congress added the building of ecological civilization and incorporated it into the overall layout of the "five-sphere integrated plan." Giving ecological civilization the status of a national strategy means to elevate development from being sustainable to being green, leaving more ecological assets for the posterity.

Since the Eighteenth National Congress, the CPC Central Committee with Comrade Xi Jinping at the core has successively put forward many important concepts such as the Belt and Road Initiative, New Normal, "supply-side structural reform," "cutting overcapacity, reducing excess inventory, deleveraging debt, lowering costs and strengthening areas of weakness," "poverty alleviation" and "Internet +," which have defined the goals and directions of China's economic reform and development and laid a solid foundation for building a moderately prosperous society in an all-round way and realizing the Chinese dream of the great rejuvenation of the Chinese nation.

At present, China has embarked on a road with Chinese characteristics of new industrialization, informatization, urbanization and agricultural modernization, for example, by promoting the further integration of informatization and industrialization, the benign interaction between industrialization and urbanization, the coordination between urbanization and agricultural modernization, and the synchronized development of industrialization, informatization, urbanization and agricultural modernization.

Since the Twelfth Five-Year Plan period, China has resolutely waged a war on pollution, sparing no effort to promote the prevention and control of atmospheric, water and soil pollution, the protection of ecological environment, with tangible results achieved in the improvement of environment quality. During the Thirteenth Five-Year Plan period, the core tasks are to improve environmental quality, enhance comprehensive environmental governance, and make up for the weak links in ecological conservation in an accelerated way.

In November 2016, the State Council issued the *Thirteenth Five-Year Plan for Ecological and Environmental Protection* (hereinafter referred to as the *Thirteenth Five-Year Environmental Plan*), which coordinates and deploys the overall environmental work during the Thirteenth Five-Year Plan period with the improvement of environmental quality at its core. The *Thirteenth Five-Year Environmental Plan* put forward the goal of achieving overall improvement of ecological environment quality by 2020 and identified seven major tasks including the three campaigns to combat atmospheric, water and soil pollution. It also stipulated twelve restrictive indicators, highlighting the systematic connection between the improvement of environmental quality and the total emission reduction, ecological conservation, the prevention and control of environmental risks and so forth.

It took the improvement of environmental quality as the core evaluation criterion, splitting the governance objectives and tasks to different regions, river basins, cities and controlled units with a list of indicators for assessment.

The *Thirteenth Five-Year Environmental Plan* demonstrates new characteristics. First, its title was changed from "environmental protection" to "ecological and environmental protection," with the content of planning being to integrate environmental protection with ecological conservation and construction. Second, in terms of philosophy of planning, it implemented the concept of environmental quality management by putting the improvement of environmental quality at its core and transforming the roadmap of the three major plans into construction drawings. When it comes to task design, it strengthened the guidance through function zoning and classification. For example, the aquatic environment in China is divided into 1,784 controlled units, 346 of which had their objectives and requirements specified one by one as they exceeded the upper limits. As for the Beijing-Tianjin-Hebei region, the Yangtze River Delta and the Pearl River Delta, the objectives and tasks of atmospheric improvement were raised and split to each category of pollutants. It set green development and reform as an important mission, changing the previous practice of taking planning as a guarantee system. It significantly strengthened the linkage between green development and ecological conservation and insisted to address the ecological and environment problems from the most fundamental aspects of development. In addition, it put forward dozens of important policies and system reform plans to reinforce the simultaneous implementation of the plan and the advancement of reform.

The State Council has reduced the number of plans in various fields and begun to pursue their quality and operability instead. In addition to the national environmental protection plan, three other plans were issued by the State Council, namely, the *Action Plan on Prevention and Control of Water Pollution* (hereinafter referred to as the *Water Action Plan*), the *Action Plan on Prevention and Control of Soil Pollution* (hereinafter referred to as the *Soil Action Plan*), and the *Comprehensive Work Plan on Energy Conservation and Emission Reduction during the Thirteenth Five-Year Plan Period*. The *Water Action Plan*, *Soil Action Plan* and the *Action Plan on Prevention and Control of Air Pollution* (hereinafter referred to as the *Air Action Plan*) which was issued in 2013 constitute the most important documents that the CPC Central Committee and the State Council have used to declare war on pollution since the Eighteenth National Congress. All the above-mentioned documents have been reviewed at the meetings of the Political Bureau before being issued by the State Council; thus, they are looked upon as the guidelines in the field of ecological and environmental protection and planning.

7. PROSPECTS FOR FUTURE DEVELOPMENT

Since the Thirteenth Five-Year Plan period, especially since the Nineteenth CPC National Congress, the CPC Central Committee with Comrade Xi Jinping at its core has promoted the eco-environment and the building of ecological civilization to an unprecedented height. The report delivered by President Xi to the Nineteenth CPC National Congress clearly stated that by 2035, "there is a fundamental improvement in the environment; the goal of building a Beautiful China is basically attained" and "in the second stage from 2035 to the middle of the twenty-first century, we will develop China into a great modern socialist country that is prosperous, strong, democratic, culturally advanced, harmonious, and beautiful." The CPC Central Committee issued the *Plan on Deepening Reform of Party and State Institutions* and established the Ministry of Ecology and Environment. The *Plan* adhered to the basic state policy of environmental protection and stressed that ecological environment be treated as life, with the implementation of the most stringent eco-environmental protection systems and the shaping of green production modes and lifestyles and with special efforts made on striking environmental problems. The *Plan* requested to integrate the duties of environment protection, which used to be scattered among different stakeholders so that the duties of supervision and administrative law enforcement of various types of pollution and emission in urban and rural areas could be executed in a unified and centralized way. It requested to strengthen the prevention and control of environmental pollution to guarantee national ecological security for the building of a Beautiful China, and it clearly laid down that the preparation and implementation of ecological and environment planning is the main responsibility of the Ministry of Ecology and Environment. The National Conference on Ecological and Environmental Protection stressed the need to strengthen eco-environmental protection in an all-round way, resolutely combat pollution, and make up for the weak links in building a moderately prosperous society in an all-round way, so as to lay the foundation for the realization of the second centennial goal. During this period of time, ground-breaking changes have taken place in ecological and environmental protection, depicting a new direction for future efforts on ecological and environmental protection planning.

7.1. China will establish a basic system to promote ecological conservation and environmental protection in a coordinated way under the guidance of ecological and environmental planning

The report of the Nineteenth CPC National Congress made strategic arrangements for environmental protection and the building of ecological conservation, providing the basic rules for ecological conservation and environmental protection planning. According to the strategic deployment for the building of ecological civilization in the new era, the ecological and environmental planning shall follow the new development concepts and target the outstanding eco-environmental problems, with the improvement of ecological and environmental quality as the core mission and exploring new boundaries and key tasks for environmental planning. At the forty-first collective learning session of the Political Bureau, General Secretary Xi Jinping delivered an important speech calling for the promotion of ecological conservation and environmental protection in an all-round, all-territory and whole-process manner so as to accelerate the formation of spatial layout, industrial structure, production mode and lifestyle conducive to resource conservation and environmental protection. He required that the country's ecological and environmental protection be planned in the whole process of socialist modernization drive, in the whole process of the building of ecological conservation and Beautiful China, and in the whole process of ecological and environment protection and governance. The institution reform for the ecological conservation is still being advanced in depth, with the relevant systems, mechanisms and policies still being improved. Therefore, the ecological and environmental protection planning should give full play to the leadership role of our institution and system, and push forward the reform through planning and the planning through reform, so as to perfect our system and policies in practice.

7.2. China will build an ecological and environmental planning system from the vertical and horizontal dimensions

Being systematic and integral means that ecological and environmental protection needs to extend to the side horizontally and to the bottom vertically, which literally means that horizontally, each and every factor at the same level shall take its due responsibility with nobody being omitted; and that vertically, all stakeholders, from the top-level central government to the bottom-level individuals, shall be held accountable. The institutional reform gives the competent environmental authorities the power to coordinate the management and supervision of ecological and environmental protection throughout all regions. In the future, the planning system should be designed horizontally to the side and vertically to the bottom and also horizontally and vertically combined. Horizontally, the ecological and environmental planning shall cover all the contents of ecological environment protection, inter alia, land, sea, mountains, rivers, lakes, forests, grasslands, urban and rural areas, main polluters, the process of pollution discharge, all environmental media and all types of pollutants so as to coordinate the ecological and environmental planning, protection, governance and supervision. We will change the previous practice of "top heavy and feet light," which means, in this case, the central government attaches great importance to environmental protection but the units at lower levels choose to respond with little actions. Vertically, we will establish an ecological and environmental planning system that goes throughout the national, provincial, municipal and county levels, with focus laid on the implementation at the municipal and county levels.

7.3. China will strengthen ecological and environmental planning by taking environment quality as the core and the space control as the focus

The eco-environment in the space of national land varies greatly, demonstrating natural characteristics unique to regions and rivers basins per se. Therefore, ecological and environmental planning shall highlight the regional and spatial attributes and systematically determine the basic framework of environment protection for key regions and the whole country alike. Thus, it is necessary to identify and differentiate the environmental features of each region, field and category, and lay stresses on the objectives and strategic missions of environment protection in general and in different stages so that we can form a planning system that has the improvement of ecological and environment quality as the core, the spatial control as the focus, and the implementation exclusive to each region, river basin and different stage. In this way, we can set up a national strategic framework for ecological and environmental planning as well as a planning system that combines key regions, river basins, fields and major policies. Currently, we have preliminarily established a space planning system for the prevention and control of water pollution based on the seven major river basins and 1,784 controlled units. We have also exploratorily specified a national atmospheric environment planning system based on three major fields, three areas (namely, the Beijing-Tianjin-Hebei region, the Yangtze River Delta and the Pearl River Delta) and ten city clusters. Yet, we still need to further explore the regionalized planning system for other factors and fields.

7.4. China will establish a complete planning system that covers the whole process of planning, including preparation, implementation, evaluation, assessment and supervision, by improving the full-chain management system of ecological and environmental planning

Whether a plan can succeed or not is highly dependent on the system and mechanism in which it is formulated and implemented. A complete plan should contain the whole links from formulation, implementation, supervision and evaluation to the accountability system. In the preparation of plans, all stakeholders should participate in the decision-making related to environmental planning, where opinions and suggestions on the planning shall be solicited from government agencies at all levels, the public and relevant polluters. Moreover, when it comes to the implementation of the plan, it is still necessary to specify who is the leading institution and who is the supporting one. The responsibilities of all stakeholders shall be clearly defined so as to avoid the overlapping or absence of responsibilities. For example, in the planning and policy formulation for the prevention and control of water pollution, the ecological and environmental authorities should play the leading role while other departments of water conservancy, transportation, agriculture and so forth should play the supporting role. During the implementation of the plan, we must make detailed schedules or action plans with clear objectives, deadlines and tasks defined on a basis of monthly or semi-annual scheduling mechanism so that we can have a better command of the implementation progress on a timely basis to address the difficulties arising in the execution as soon as possible. As far as the assessment of environmental planning is concerned, we shall comprehensively monitor the assessment and promote the implementation of the plan by establishing an annual assessment mechanism and a follow-up assessment mechanism. Last but not least, we should launch a corresponding administrative accountability system to ensure that it serves as the basis for the evaluation of officials' performance, for the purpose of urging local officials to pay close attention to environmental protection, the implementation of planning and to provide assistance in the fulfillment of planning objectives.

7.5. China will establish a supporting system adapted to the development of ecological and environmental planning from the perspectives of theories, methods, technical tools, institutional teams and market mechanisms

Future efforts shall be spent mainly on the problems of resources, environment and ecology arising in the rapid urbanization and industrialization, especially in response to the people's needs of higher standard of living that have appeared in China's building of a moderately prosperous society in an all-round way and in its socialist modernization drive. Currently, the researchers of environmental planning in our country have maintained their due sensitivity to the new technology abroad, and are also brave enough to make innovations in the application of the technologies. Relevant technical methods of environmental evaluation, simulation and prediction in certain fields have been relatively developed to maturity.[15] In the future, we should strengthen the research on technical methods where we are weak now, such as environmental optimization and integration, and develop more applicable and targeted methods tailored to the characteristics of different fields. In addition, we shall let the environmental planning take the leading role by further developing environmental planning research institutions, facilitating the teaching of environmental planning theory in colleges and universities, encouraging more talents to engage in the preparation and implementation of environmental planning and fully mobilizing the initiative of various localities to prepare their own environmental plans.

8. CONCLUSION

Since China adopted the reform and opening-up policy forty years ago, the dialectical relationship between development and protection has been continuously adjusted in line with the different stages of the country's economic and social development, pushing the modification of the system and mechanism of ecological and environmental protection on a constant basis. The environmental planning has been developing in harmony with economy and society in a complementary and spiral manner, with tangible progress and successful experience.

China's economic growth has changed from pursing high-speed to high-quality at the present stage, a critical period featuring the transformation of development modes, the optimization of economic structures and the conversion of growth engines. The environment in the stage of high-quality development can no longer be regarded as a valueless or low-cost factor of production, nor can it be looked upon only as a condition to back development. Instead, it should be regarded as a scarce resource that deserves intensive protection and repair. This is the shift in the positioning of ecological and environmental protection as times go by. Environmental protection planning in the new era should focus on improving the environmental quality, promote ecological conservation and environmental protection in a coordinated manner, and strengthen management and support to facilitate implementation. We should be bold in innovation, strive to improve the environmental governance system and modernize the governance capacity to build an all-dimensional planning system for green development and blaze a trail for China's modernization featuring efficient production, productive life and good ecology.

Table 3-2 List of Key Eco-environmental Protection Plans since the Reform and Opening-up

Periods		Plans	Competent authorities
Fifth Five-Year Plan Period (1975–1980)		China's Ten-Year Plan for Environmental Protection	Leading Team for Environmental Protection under the State Council
Sixth Five-Year Plan Period (1980–1985)		Special chapters were included in the Sixth Five-Year Plan and no separate plans were issued.	NHFPC
Seventh Five-Year Plan Period (1986–1990)	1987	China's Ten-Year Plan for Environmental Protection (1986–1990)	NHFPC, Environmental Protection Commission of the State Council
	1989	Outline of China's Environmental Protection Plan in 2000	Adopted at the Third National Environmental Protection Conference
Eighth Five-Year Plan Period (1991–1995)	1992	China's Ten-Year Plan for Environmental Protection and Outline of the Eighth Five-Year Plan	National Environmental Protection Agency
	1992	China's Action Plan on Environmental Protection (1991–2000)	National Environmental Protection Agency
	1992	National Ten-Year Plan on the Comprehensive Improvement of the Urban Environment and the Eighth Five-Year Plan	National Environmental Protection Agency
	1990	National Ten-Year Plan on the Protection of Nature Reserves and Species and the Eighth Five-Year Plan	National Environmental Protection Agency
	1990	National Ten-Year Plan on the Protection of Marine Environment and the Eighth Five-Year Plan	National Environmental Protection Agency

Table 3-2 Continued

Periods		Plans	Competent authorities
	1990	National Ten-Year Plan on the Management of Radioactive Environment and the Eighth Five-Year Plan	National Environmental Protection Agency
	1990	China's Eighth Five-Year Plan and Outline of the Ten-Year Plan on the Prevention and Control of Environmental Noise and the	National Environmental Protection Agency
Ninth Five-Year Plan Period (1996–2000)	1996	China's Ninth Five-Year Plan on Environmental Protection and Long-Range Objectives Through the Year 2010	For the first time approved by the State Council
	1996	National Plan for Controlling the Total Emission of Major Pollutants During the Ninth Five-Year Plan Period	Appendix to the Ninth Five-Year Plan
	1996	China Trans-Century Green Project Plan (Phase One)	Appendix to the Ninth Five-Year Plan
	1998	National Program of Ecological Environmental Construction	Issued by the State Council
	1996	Plan on the Prevention and Control of Water Pollution and the Ninth Five-Year Plan in the Huaihe River Basin	Approved by the State Council
	1998	The Ninth Five-Year Plan and the Plan Through the Year 2010 on the Prevention and Control of Water Pollution in Taihu Lake	Approved by the State Council
	1998	The Ninth Five-Year Plan and the Plan Through the Year 2010 on the Prevention and Control of Water Pollution in Dianchi Lake	Approved by the State Council

(continued)

Table 3-2 Continued

Periods		Plans	Competent authorities
	1999	The Plan on the Prevention and Control of Water Pollution in the Haihe River Basin	Approved by the State Council
	1999	The Ninth Five-Year Plan and the Plan Through the Year 2010 on the Prevention and Control of Water Pollution in Liaohe River Basin	Approved by the State Council
	1994	Action Plan for the Conservation of Biodiversity	13 ministries including National Environmental Protection Agency and NHFPC
	1997	National Outline of the Nature Reserve Development Plan (1996–2010)	National Environmental Protection Agency, NHFPC
Tenth Five-Year Plan Period (2001–2005)	2001	China's Tenth Five-Year Plan on Environmental Protection	Approved by the State Council
	2001	National Plan for Controlling the Total Emission of Pollutants During the Tenth Five-Year Plan Period	Appendix to the Tenth Five-Year Plan
	2002	China's Key Project Plans on Environmental Protection during the Tenth Five-Year Plan	Appendix to the Tenth Five-Year Plan
	2001	Blue Sea Action Plan for the Bohai Sea (2001–2015)	Approved by the State Council
	2000	The Outline of China's Environmental Protection	Issued by the State Council
	2001	the Tenth Five-Year Plan on the Prevention and Control of Water Pollution in Taihu Lake	Approved by the State Council
	2002	the Tenth Five-Year Plan on the Prevention and Control of Water Pollution in Chaohu Lake	Approved by the State Council

Table 3-2 Continued

Periods		Plans	Competent authorities
	2003	the Tenth Five-Year Plan on the Prevention and Control of Water Pollution in Dianchi Lake	Approved by the State Council
	2003	the Tenth Five-Year Plan on the Prevention and Control of Water Pollution in the Huaihe River Basin	Approved by the State Council
	2003	the Tenth Five-Year Plan on the Prevention and Control of Water Pollution in the Haihe River Basin	Approved by the State Council
	2003	the Tenth Five-Year Plan on the Prevention and Control of Water Pollution in the Liaohe River Basin	Approved by the State Council
	2003	Plan on the Pollution Control in the Eastern Route of the South-to-North Water Diversion Project (2001–2010)	Approved by the State Council
	2001	Plan for the Prevention and Control of Water Pollution of the Three Gorges Reservoir and Its Upper Reaches (2001–2010)	State Environmental Protection Administration (SEPA)
	2002	Tenth Five-Year Plan for Prevention and Control of Acid Rain and Sulfur Dioxide Pollution in the Two Control Zones	SEPA
	2004	National Plan for Development of Hazardous Waste and Medical Waste Disposal Facilities	SEPA
	2001	SEPA's Fourth Five-Year Plan on the Publicity and Education About Environmental Laws	SEPA
11th Five-Year Plan Period (2006–2010)	2007	China's 11th Five-Year Plan on Environmental Protection	Issued by the State Council

(continued)

Table 3-2 Continued

Periods		Plans	Competent authorities
	2006	National Plan for Controlling the Total Emission of Major Pollutants During the 11th Five-Year Plan Period	Officially replied by the State Council
	2006	The 11th Five-Year Plan for the Prevention and Control of Water Pollution in the Songhuajiang River Basin	Approved by the State Council
	2008	Master Plan for Comprehensive Control of Water Environment of the Taihu Lake Basin	Approved by the State Council
	2006	Water Pollution Control and Soil and Water Conservation Planning in Danjiangkou Reservoir Area and Its Upper reaches	Approved by the State Council
	2008	Plan for the Prevention and Control of Water Pollution in the Huaihe River Basin (2006–2010)	Ministry of Environmental Protection (MEP), NDRC, Ministry of Water Resources (MWR), and Ministry of Housing and Urban-Rural Development (MOHURD)
	2008	Plan for the Prevention and Control of Water Pollution in the Haihe River Basin (2006–2010)	MEP, NDRC, MWR, MOHURD
	2008	Plan for the Prevention and Control of Water Pollution in the Liaohe River Basin (2006–2010)	MEP, NDRC, MWR, MOHURD
	2008	Plan for the Prevention and Control of Water Pollution in the Chaohu River Basin (2006–2010)	MEP, NDRC, MWR, MOHURD

Table 3-2 Continued

Periods		Plans	Competent authorities
11th Five-Year Plan Period (2006–2010)	2008	Plan for the Prevention and Control of Water Pollution in the Dianchi Lake Basin (2006–2010)	MEP, NDRC, MWR, and MOHURD
	2008	Plan for the Prevention and Control of Water Pollution in Middle and Upper Reaches of the Yellow River (2006–2010)	MEP, NDRC, MWR, MOHURD
	2008	Plan for the Prevention and Control of Water Pollution of the Three Gorges Reservoir and Its Upper Reaches	MEP
	2006	Action plan of Building Moderately Prosperous Rural China	SEPA
	2006	The National 11th Five-Year Plan for Eco-Environmental Protection	SEPA
	2007	Outline of National Plan for Key Ecological Function Areas	SEPA
	2007	Outline of National Plan for Protection and Use of Biological Species Resources	SEPA
	2007	Outline of National Plan for Prevention and Control of Environmental Pollution in Rural Areas (2007–2020)	SEPA
	2008	11th Five-Year Plan for National Prevention and Control of Acid Rain and Sulfur Dioxide	SEPA
	2006	11th Five-Year Plan for National Environmental Scientific and Technological Development	SEPA

(continued)

Table 3-2 Continued

Periods		Plans	Competent authorities
	2007	Special Program of National Key Laboratory on Environmental Protection Under the 11th Five-Year Plan	SEPA
	2007	Special Program of National Environmental Protection Technical Center Under the 11th Five-Year Plan	SEPA
	2007	Construction Plan for the National Environmental Technology Management System	SEPA
	2006	China's 11th Five-Year Plan on Environmental Protection Standards	SEPA
	2007	National Action Plan for the Environment and Health 2007–2015	Ministry of Health
	2005	Development Plan for National Environmental Laws and Regulations in the 11th Five-Year Plan	SEPA
	2006	SEPA's Fifth Five-Year Plan on the Publicity and Education About Environmental Laws	SEPA
	2008	11th Five-Year Plan for Developing National Environmental Supervising Capacity	NDRC and MOF
	2007	11th Five-Year Plan for Developing Harmless Disposal Facilities for Urban Domestic Waste	NDRC
	2007	11th Five-Year Plan for Prevention and Control of Sulfur Dioxide in Existing Coal-Burning Power plants	NDRC
	2007	11th Five-Year Plan for Energy Development	NDRC

Table 3-2 Continued

Periods		Plans	Competent authorities
	2005	Program on Integrated Treatment of Chrome Slag Pollution	NDRC and SEPA
12th Five-Year Plan Period (2011–2015)	2011	National 12th Five-Year Plan for Environmental Protection	Issued by the State Council
	2012	Comprehensive Program on Energy Saving and Emission Reduction During the "12th Five-Year Plan" Period	Issued by the State Council
	2012	Plan for Developing Energy-Saving and Environmental Protection Industries During the 12th Five-Year Plan Period	Issued by the State Council
	2011	12th Five-Year Plan for Comprehensive Prevention and Control of Heavy Metal Pollution	Approved by the State Council
	2011	Plan for Prevention and Control of Water Pollution of the Mid and Lower Reaches of the Yangtze River (2011–2015)	Approved by the State Council
	2011	National Plan for Prevention and Control of Groundwater Pollution (2011–2015)	Approved by the State Council
	2012	Plan for Prevention and Control of Water Pollution in Key River Basins (2011–2015)	Approved by the State Council
12th Five-Year Plan Period (2011–2015)	2012	12th Five-Year Plan for Water Pollution Control and Soil and Water Conservation in Danjiangkou Reservoir Area and Its Upper reaches	Approved by the State Council
	2012	12th Five-Year Plan for Prevention and Control of Atmospheric Pollution in Key Regions	Approved by the State Council

(continued)

Table 3-2 Continued

Periods		Plans	Competent authorities
	2012	12th Five-Year Plan for Nuclear Safety and Prevention and Control of Radioactive Pollution & Long-range Objectives for 2020	Approved by the State Council
	2012	12th Five-Year Plan for Developing Harmless Disposal Facilities for Urban Domestic Waste	General Office of the State Council
	2012	12th Five-Year Plan for Developing Treatment and Recycling Facilities for Urban Sewage	General Office of the State Council
	2010	Strategy and Action Plan in China for Biological Diversity Protection (2011– 2030)	MEP
	2011	National Mid and Long-Term Plan for Developing Talents in Eco-Environmental Protection (2010–2020)	MEP
	2011	12th Five-Year Plan for National Environmental Scientific and Technological Development	MEP
	2011	The 12th Five-Year Plan on Environmental Protection and Work Plan on Environment and Health for the 12th Five-Year Plan Period	MEP
	2011	12th Five-Year Plan for National Environmental Monitoring	MEP
	2011	12th Five-Year Plan for Developing National Environmental Laws and Regulations and Environmental Economic Policies	MEP
	2011	12th Five-Year Plan for Environmental Impact Assessment	MEP

Table 3-2 Continued

Periods		Plans	Competent authorities
	2012	12th Five-Year Plan for Prevention and Control of Pollution from Persistent Organic Pollutants in Major Industries	MEP
	2012	12th Five-Year Plan for National Integrated Rural Environment Management	MEP
	2012	Twelfth Five-Year Plan for Prevention and Control of Pollution by Hazardous Waste	MEP
	2011	National 12th Five-Year Plan for Prevention and Control of Pollution in Livestock and Poultry Breeding	MEP
	2013	National 12th Five-Year Plan for Eco-Environmental Protection	MEP
	2013	12th Five-Year Plan for Prevention and Control of Environmental Risks of Chemicals	MEP
	2013	China's 12th Five-Year Plan on Environmental Protection Standards	MEP
	2013	Program on the Implementation of International Environmental Conventions in the 12th Five-Year Plan Period	MEP
	2013	12th Five-Year Plan for Developing National Environmental Supervising Capacity	MEP
	2014	The National Plan for Eco-Environmental Protection and Construction (2013–2020)	MEP and other authorities

(continued)

Table 3-2 Continued

Periods		Plans	Competent authorities
	2011	Plan for Ensuring Hygiene and Safety of Urban Drinking Water Nationwide (2011–2020)	MEP and other authorities
	2011	12th Five-Year Plan for National Energy Technologies (2011–2015)	MEP and other authorities
	2012	Special Program for Waste Recycling Projects Under the 12th Five-Year Plan	MEP and other authorities
	2013	Action Plan for Prevention and Control of Air Pollution	Issued by the State Council
	2015	Action Plan for Prevention and Control of Water Pollution	Issued by the State Council
	2015	Eco-Environmental Protection Plan on the Coordinated Development of the Beijing-Tianjin-Hebei Region	MEP and other authorities
13th Five-Year Plan Period (2016–Present)	2016	Eco-Environmental Protection Plan during the 13th Five-Year Plan	Issued by the State Council
	2016	Action Plan for Soil Pollution Prevention and Control	Issued by the State Council
	2016	Comprehensive Program on Energy Saving and Emission Reduction During the 13th Five-Year Plan Period	Issued by the State Council
	2017	Plan for Prevention and Control of Water Pollution in Key River Basins (2016–2020)	MEP and other authorities
	2016	The National 13th Five-Year Plan for Eco-Environmental Protection	MEP and other authorities
	2016	National Urban Eco-Environmental Protection and Development Plan During the 13th Five-Year Plan Period (2015–2020)	MEP and other authorities

Table 3-2 Continued

Periods		Plans	Competent authorities
	2016	13th Five-Year Plan for National Environmental Scientific and Technological Development	MEP and other authorities
	2017	13th Five-Year Plan for National Integrated Rural Environment Management	MEP and other authorities
	2017	The 13th Five-Year Plan on Environmental Protection and Work Plan on Environment and Health for the 13th Five-Year Plan Period	MEP and other authorities
	2017	13th Five-Year Plan for Nuclear Safety and Prevention and Control of Radioactive Pollution & Long-range Objectives for 2025	MEP and other authorities
	2017	China's 13th Five-Year Plan on Environmental Protection Standards	MEP and other authorities
	2016	Plan for Nationwide Environmental Protection Authorities to Provide Paired Assistance to Xinjiang During the 13th Plan Period	MEP
	2016	Plan for Nationwide Environmental Protection Authorities to Provide Paired Assistance to Tibet During the 13th Plan Period	MEP
	2017	Eco-Environmental Protection Plan on the Yangtze River Economic Belt	MEP and other authorities
	2017	The Belt and Road Ecological and Environmental Cooperation Plan	MEP

Notes

1. Li Xiaoxi and Lin Yongsheng, 2017, The Development of China's Market Economy in the Past Forty years of Reform and Opening-up, *Globalization*, 7th Edition, pp. 55–66.
2. Ren Junhong (2015), The Historical Status of China's First Environmental Protection Conference", *Journal of Hunan University of Administration*, 1st Edition, pp. 124–128.
3. Zhou Jun, Ni Yanfang, Xing Jia, Zhang Li, Teng Zhikun and He Pingping (2013). An Analysis of The Development Trend and Existing Problems of Environmental Planning in China. *Environmental Science and Management*, 4th Edition, pp. 185–187.
4. The previous five meetings were all named "National Environmental Protection Conference", and the sixth meeting was officially renamed "National Environmental Protection Conference" since 2006.
5. Li Zhengqiang, Shu Hong (2008). Economic development and environmental Protection, *Theory and Reform*, No. 3, pp. 65–66.
6. Wang Jinnan, Su Jieqiong and Wan Jun (2017). Theoretical Connotation and Realization Mechanism Innovation of "Clear Water and Green Mountains are Gold and Silver Mountains", *Environmental Protection*, No. 11, pp. 13–17.
7. Zou Shoumin, Wu Shunze and Xu Yi, et al (2008), "Review, analysis and prospect of national environmental protection planning", paper presented at the 2008 annual conference of environmental planning Committee, Chinese Society for Environmental Sciences, p. 309.
8. Sun Rongqing (2012). The Development Course of the Five-year Environmental Protection Plan, *China Environment News*, August 9, p. 2.
9. Guo Xiaomin (1993). Review and Prospect of China's Environmental Planning, *Environmental Science*, No. 4, pp. 10–15.
10. Mu Shilong (2016). Exploration of Economic System Reform after Reform and Opening up, Master's thesis of Liaoning University of Technology, p. 11.
11. Li Shantong, Wu Sanmang, He Jianwu and Liu Ming (2012). Review and Prospect of China's Economic Development in the Decade after China's Accession to the WTO, *Journal of Beijing Institute of Technology*, No. 3, pp. 1–7.
12. Yao Jingyuan (2011). Achievements, Problems and Prospects in 10 Years after China's Accession to the WTO, *Red Flag Manuscript*, No. 15, pp. 21–24.
13. Notice of the State Council on Printing and Distributing the National Environmental Protection "12th Five-Year Plan", Guo Fa [2011] No. 42.
14. Circular of the State Council on Printing and Distributing the 13th Five-Year Plan for Ecological and Environmental Protection, Guo Fa [2016] No. 65.
15. Yan Xiaopin, Zhang Zhenzhen and Liu Yong (2013). Assessment and Analysis of the current Situation of Environmental Planning Technology in China, *Environmental Pollution Prevention & Control*, No. 4, pp. 104–109.

CHAPTER FOUR

Circular Economy

QI JIANGUO, WANG YINGJIE AND MA XIAOQIN[*]

1. INTRODUCTION

In 1978 when China launched its reform and opening-up initiative, the country was still largely an agricultural economy, with more than 80% of the population living in the countryside. The per capita GDP was RMB381, or USD 226 at the exchange rate then. The per capita disposable income stood at RMB343 for urban residents and RMB133 for their rural counterparts. The per capita consumption level of material products stayed very low, with the overall living standard not high enough to meet the most basic needs. In those days, lucid water and lush mountains could not provide people with enough food or clothes.

The reform and opening-up have brought not just the development of economy. It also means an increasing amount of material products consumed and of wastes from industrial production. The quantity of wastes from, inter alia, primary, secondary and tertiary industries, households, textiles, kitchens and packages has increased by twenty to thirty times. Compared with 1978, wastes are

---

[*] Qi Jianguo is a researcher at the Institute of Quantitative and Technical Economics at the Chinese Academy of Social Sciences and chief policy expert at the China Association for Circular Economy; Wang Yingjie is a doctoral student at the Graduate School of the Chinese Academy of Social Sciences; Ma Xiaoqin is a doctoral student at the Graduate School of the Chinese Academy of Social Sciences.

increasing in categories, with many of them unheard of in 1978, such as used mobile phones, rechargeable batteries and lithium-ion batteries, as well as various kinds of discarded electronic devices. However, in the past, the disposal of such solid waste was mainly through landfill, which literally besieging many large and medium-sized cities with wastes. The situation of waste water, which was mostly randomly discharged, was no better: completely polluted surface water, eutrophicated lakes, black and muddy rivers, and poisonous and smelly underground water. Smog haunted us due to exhaust emissions. The lucid water and lush mountains became scarred and battered. Economic growth was in the middle of a paradox: why should we develop the economy? For happiness, some might say. Are we happier when the economy has grown by thirty times? No is the answer because the environment is getting worse and the ecology is constantly under destruction. Poor if economy does not grow and foul if it does.

A serious problem is blocking the wheels of the twenty-first century: how can we find a way to keep the economy growing rapidly while improving the ecological environment alongside? Or in other words, how can we turn lucid water and lush mountains into invaluable assets, while at the same time let our people enjoy a good life?

The answer is to vigorously develop circular economy and ecological conservation.

2. THE DEVELOPMENT ORIENTATION OF CIRCULAR ECONOMY

2.1. Classical circular economy

The classical circular economy and production have long existed. The typical example was the Mulberry-fish Pond model as early as more 1,000 years ago in southern China. At that time in Pearl River Delta area, in order to take full advantage of land, people used the sludge dug from the pond to plant mulberry trees and pond itself to raise fish - a closed yet efficient artificial ecosystem that could drive the growth of both fishing and silk industry. The mulberry leaves were the food of silk worms, whose feces were the feed for fish, whose feces sank to the riverbed and became sludge that would be used as fertilizer for mulberry trees. Thus, a closed cycle of ecotype economy was formed. The result was self-evident: more mulberry leaves, more silkworms, more silkworm feces and naturally more fish. What humans got was silk and fish in larger quantities. The accompanying industry was that of silk and textile. This is the typical classical circular economy model, in which economy continued growing but without waste.

In this circle, sunlight, water, carbon dioxide and trace elements in soil constitute the basis for the cycle of matter. Fish and silkworm cocoons, as the final products, become the economic products of human beings. This idea had existed objectively before the birth of the doctrine of industrial ecology, according to whose basic principle, there is a balanced natural ecological chain between the wastes' producers and decomposers, which maintains the balance between species and material in nature. Fish is the decomposer of silkworm excreta, mulberry tree serves as the decomposer of fish excreta, and silkworm works as the decomposer of mulberry leaves.

In the same way as the natural ecological system, the production system set up by humans for economic development also produces wastes, and there are also "decomposers" which use the wastes as raw materials. However, unlike natural ecosystems, if the decomposers of wastes cannot obtain economic benefits, they will disappear, which means that waste that is not disposed of will increase until the amount of waste accumulated causes environment pollution which threatens the survival of human beings. Only then can the decomposition of wastes be looked upon and addressed as an integral part of economic and social activities.

The earliest systematic analyst of the recycling of industrial wastes was Karl Marx, the great revolutionary mentor of the proletariat. He said, "By reconverting the excrements of production, the so-called offal, into new elements of production, either of the same, or of some other line of industry, these so-called excrements are thrown back into the cycle of production and consequently of consumption (whether productive or individual). It is only as excrements of combined production on a large scale that they become valuable for the productive process as bearers of new exchange-values. These excrements, aside from the services which they perform as new elements of production, reduce the cost of raw material to the extent that they are saleable. The reduction of the cost of this portion of constant capital increases to that extent the rate of profit, assuming the amount of the variable capital and the rate of surplus-value to be given quantities."[1] Obviously, Marx discussed the idea of circular economy from the perspective of saving capital and increasing profit margin, but Marx did not use the concept of circular utilization. It can be seen that, more than 100 years ago, Marx had already observed that there were "decomposers" of industrial wastes in capitalist production, but only when decomposers, or waste recyclers, could raise their profit rate from the use of waste would they do it. They do it not because they want to reduce waste emissions or environmental pollution caused by production.

By the 1960s, the industrialization in developed countries was coming to an end. Under the technical level at that time, non-renewable resources were in short supply. American economist Kenneth E. Boulding put forward the theory of "spaceman economy," which was inspired by the launch of space shuttles. Boulding believed that the Earth was a single spaceship, which could survive by consuming internal resources and would perish upon the depletion of resources. Therefore, it is necessary to continuously recycle the limited resources and reduce waste emissions so that human beings can survive for a long time.

In 1972, the Club of Rome published its well-known study report *The Limits to Growth*, which demonstrates that there are limits to economic growth due to limited resources, reminding people to save resources.

It can be seen from the above analysis that people still viewed the consequences of resource utilization mainly from the perspectives of the limitation of resources and the economic efficiency of resource utilization in the early 1970s. Whether it is improving resource utilization efficiency or recycling waste, they started from the viewpoint of the sustainability of resource supply and economic efficiency of resource utilization. For that reason, this type of classical circular economy is also called the resource-oriented circular economy. But at this stage, people paid little attention to the effects of resource utilization on environment.

It was not until the early 1970s when the environmental pollution in developed countries became increasingly serious and when the scholars summarized the previous eight famous pollution accidents did human beings begin to pay attention to environmental problems. However, the focus of attention at that time was still on the end-treatment, i.e. environmentally sound disposal of discharged pollution, with less efforts on preventing the generation of pollutants from the source.

2.2. Modern circular economy

Modern circular economy refers to the circular economy mode in which developed countries managed wastes with the goal of environmental protection when they had completed industrialization in the 1970s. In his book *The Blueprint of a Green Economy* published in 1989, the British scholar David Pearce used the term "circular economy" for the first time. A year later, in the book *Economics of Natural Resources and the Environment* co-authored by David Pearce and Kerry Turner, circular economy was used to define the recycling of resources. Although they still equaled circular economy to resource recycling, they studied it from the perspective of environmental protection.

Germany is the first country in the world to develop circular economy through the comprehensive utilization of wastes. Germany is also one of the earliest countries to promulgate laws on the safe disposal of wastes and on environmental protection. In 1994, Germany officially drafted the *Closed Substance Cycle and Waste Management Act*, which came into effect in 1996. It is the first national law on circular economy globally. Then, in 2000, Japan formulated the *Basic Act for Establishing a Sound Material-Cycle Society*. These two laws pioneered the legislation for circular economy at national level. Subsequently, a large number of countries followed by developing different types of laws and regulations for the recovery and comprehensive utilization of wastes. This indicates that shift of the development of circular economy from being resource-oriented to environment-oriented. We call it the modern circular economy.

Many Chinese scholars have defined modern circular economy from perspectives like philosophy, resource utilization, environmental protection, material flow, ecology and technology paradigm. By integrating these definitions, we can infer that the modern circular economy is, in nature, an ecotype economic development model. It is the result of increasingly deep public understanding of the contradictions between economic growth, resource constraints and environmental pollution. Its aim is to improve the utilization rate of resources and to protect the environment through saving resources and recycling wastes. It is supported by innovations in technology and driven by breakthroughs in system. Under the premise of being technically feasible, economically reasonable and meeting the needs of the society and market, circular economy realizes sustainable growth by minimizing waste discharge and maximizing resource efficiency.

At the technical level, the modern circular economy emphasizes, in the first place, the use of productive technologies capable of efficiently utilizing resources and saving resources by reducing consumption per unit output. Second, by combining the technologies for production and environmental protection, it forms a clean production system where pollutants generated and discharged in the production process will be reduced, even to "zero" in some cases. Third, it will realize the cycle use of material resources through the comprehensive recycling and reuse of wastes. Last, it achieves ecological balance and sustainable development of economy and society by the environmentally sound disposal of wastes.

2.3. The theoretical basis of researches on circular economy

Theory comes from practice, and in turn guides practice. The theory and thoughts of the circular economy have formed and developed as a result of human's long-term practice. Circular economy has a rich content; therefore, its theory is inevitably interdisciplinary. Viewed from the current academic research results, it involves ecology, ecological economics, resource economics, environmental economics, technical economics, institutional economics, engineering science, system science, collaborative science and so on.

2.3.1. Circular economy and the theory of ecology

Ecology was, at the very beginning, a subject of animal research and later it developed into the study on ecosystem. Ecosystem is the organic complex of living organisms, their physical environment, and all their interrelationships with the energy flow, material flow, information flow, and value flow in a particular unit of space. Ecosystem is an ever-evolving dynamic system.[3]

Ecological balance can be defined as a state of stable operation of ecosystem. In the course of maintaining ecological balance, human and the environment on which they live interact with and influence each other. The history of human civilization is also a history of development or a history of ecological evolution, in which humans compete and coexist with natural and social environment. The contradiction between human, resources and environment, in essence, is the imbalance of inter-relationship in the ecosystem. As a member of the ecosystem, human beings are not only restricted by the general natural law but also the most active and dynamic factor that dominates the ecosystem. Once recognizing and mastering the characteristics of ecosystems, human beings can manage them scientifically, prevent their reverse evolution, create new systems with higher ecological and economic benefits, and achieve a new ecological balance.

Circular economy, as a development mode, emerges as a response to the serious problems of resources and environment that have threatened the sustainability of human beings. It is an innovation after much reflection on the traditional growth modes. By regarding the social economic system as a sub-system of the ecosystem, humans strike the ecological balance by seeking a new way to coordinate the operation of the economic system and the natural system according to the ecological law and by means of technology. Therefore, simple as they are, the ideas of the harmonious development of human and nature, the conversion of material energy of ecosystem and the ecological balance have become an important ideological foundation for circular economy theories.

2.3.2. Circular economy and the theory of ecological economics

Ecological economics is an interdisciplinary science on the benign operation of eco-economic system, which is formed in the interaction between ecosystem and economic system, with its research object being the general laws regarding the material exchange, energy flow, information transfer, value transfer and appreciation and the internal connections of the four aspects, which exist in and between economic system and ecosystem in the process of regeneration of social materials.

According to ecological economics, ecosystem has an ecological threshold, which means environment's self-purification and carrying capacity is not unlimited. The ecological system can be restored under the regulation of mutual feedback of economic activities when the impact of an economic system does not exceed the ecosystem threshold, or will witness imbalances, loss of control and even catastrophic consequences for the entire human race if it does. In other words, the structure of the compound system of ecological economy, its interaction and evolution in/with ecological economy, and the scale of economic activities together play a crucial role in the stability and recoverability of ecosystem. This has laid a certain foundation for the application researches of the theory of circular economy, which mainly studies the interaction between economy and environment and advocates improving and protecting the self-organizing capacity of ecosystem under the premises of respect for nature, with focus laid on the cycle use of resources, the recycling of wastes and the reduction of waste discharge to make the interference of economic activities in ecology lower than the ecological threshold and realize the win-win of economic development and environmental protection.

2.3.3. Circular economy and the theory of resource economics

Resources are the material basis for human survival and development. Resource economics and circular economy take the allocation efficiency and sustainable utilization of resources as the main content on the material level. Unlike the traditional resource economics which mainly studies natural resources, circular economy is about waste resources and renewable resources. From the perspective of sustainable development, resource economics focuses on the sustainable supply of natural resources, while circular economy pays closer attention to the environmental sustainability amongst resource utilization. When it comes to the dependence on system, the former is more sensitive to policies on resource exploitation while the latter is keener on resource policy and environmental policy.

2.3.4. Circular economy and the theory of environmental economics

Environmental economics is a discipline that has appeared when the waste that is generated in human economic and social activities approaches the ecological threshold and gives environment the attributes of resources. It believes that the environment is a kind of resources with the characteristics of scarcity, multipurposes, proliferation and difficulty in measurement. The core of researches in environmental economics is to understand and properly handle the relationship between environmental protection and economic development. Circular economy, however, is a technical way to deal with the contradictions between economic development and environment, with the aim of seeking a virtuous circle between the two. Only when the environment becomes a type of scarce resources can circular economy show its value; otherwise, it will only be a technical solution to the sustainable supply of resources. In this sense, circular economy can be described as an optimized solution jointly offered by resource economics and environmental economics.

2.3.5. Circular economy and the theory of technological economics

Technological economics studies the relationship between technology and economy and that between input and output of economic activities, that is, the comparative relationship between the costs and benefits of the technical scheme of certain economic activities. According to the principle of technological economics, circular economy must be based on the feasibility and economic viability of resource recycling technology, neither of which is dispensable. But in reality, the recycling of resources is not economically reasonable because of the low technical efficiency, which means that the so-called recycling is costly and cannot make it in the market. Against this backdrop, it is critical to research and develop more efficient technologies, which, in this context, not only refer to the hard technologies applied in production, but also the management technologies that optimize the combination of factors. Therefore, from the perspective of technological economics, circular economy means to compare and optimize technical schemes that replace new resources with wastes.

2.3.6. Circular economy and the theory of institutional economics

Unlike traditional resources, waste resources will become pollutants if not recycled and reused. This is largely determined by the basic characteristics of wastes. Therefore, the utilization of waste resources involves not only the direct costs and benefits in the utilization but also the environmental resources, which means that the recycling of waste resources has strong externalities. The externality problems generally need to be solved through institutional innovation, which means that there is an inherent link between circular economy and institutional economics.

As a new mode of economic development, circular economy, on the material level, is mainly characterized by the replacement of new resources with wastes. But whether this replacement can be sustainable depends on whether its costs and benefits have comparative advantages over that of new resources. In many cases, the core problem of utilizing waste resources is that the cost of recycling waste and converting it into useful materials is higher than that of new resources. The cost of using waste resources will not be comparatively advantageous in cost-effectiveness if the environmental cost is not taken into account, but it will if the negative externalities of new resource development and the positive externalities of waste resource utilization are included in the economic accounting.

Therefore, to develop circular economy, we need to internalize the externality of waste resource utilization through institutional innovation, which, in practice, is mainly seen in the extension of producers' responsibility and the environmental tax on discharged wastes. In essence, the extension of producers' responsibility means that the producers assume the economic obligation for the recovery and disposal of the products produced by them which have become wastes after being consumed. Proper compensation for the major stakeholders engaged in waste recycling can effectively improve the comparative advantages of and thereby encourage waste resource recycling and utilization. Taxes on waste emissions can also internalize negative externalities. To reuse waste instead of discharging it can also save the cost in emissions and encourage recycling of waste. The waste producers here can be defined as either a direct manufacturer or a consumer of the material products. No matter which party bears the waste disposal cost, waste users can save money.

3. THE COURSE OF DEVELOPMENT

The term "circular economy" did not appear in China's academia until 1998, but the practice of circular economy existed long ago. For example, the typical circular economy model in agriculture, such as the "mulberry and fishpond system" and the "duck-rice symbiosis," appeared in ancient China, but was not named "circular economy." Therefore, we can divide the development course of circular economy in China into three stages, namely, the classical stage, the transitional stage and the rising stage of modern circular economy, with each stage to be divided into several sub-stages.

3.1. The resource-oriented classical stage (before 1998)

3.1.1. The stage of classical circular economy dominated by the utilization of solid waste resources (before 1978)

Before 1978, China was still in the economic stage of agricultural dominance. Although the strategy of four modernizations (modernization of agriculture, industry, national defense and science and technology) had been put forward and industry accounted for more than 40% of the national economy, the scale of industry was still small and environmental pressure was not very pressing. On the contrary, because of the backward productivity and the insufficient supply of production resources and consumer goods, people and enterprises then had to make use of all resources available, including waste resources definitely. Before 1978, China did not have an independent environmental protection agency but an Office of the Leading Team for Environmental Protection under the State Council, which was established in 1973 and was responsible for some important matters concerning environmental management. At that time, in the city, the government generally set up an office for the utilization of three wastes (waste water, waste gas and solid waste). The recycled wastes include animal bones, waste metal and other solid wastes, with waste metal to be the most important. The

Ministry of Materials and Equipment and the National Supply and Marketing Cooperatives both had a recycling system for waste resources.

3.1.2. The stage of changing waste utilization into an approach of end treatment (1979–1997)

After the reform and opening-up policy was adopted, environmental problems gradually emerged as a result of accelerated economic growth rate, the rising consumption of resources, and the increasing production and emission of wastes. Particularly, along with the industrialization, a large number of township enterprises, which were mostly of backward technology and scattered layout, emerged like mushrooms after rain, leading to aggravating environmental pollution. In 1992, Deng Xiaoping's talks on his tour to southern China brought another round of industrialization, which signified China's entry into the stage of large-scale industrialization. After 1996, China basically ended the history of the shortage of industrial processed products and entered the "relative surplus age," but the shortage of basic raw materials became more prominent. The direct consequence of the rapid industrialization is that China came into the traditional technological and economic paradigm track of industrialization in developed countries, which means large-scale production, consumption, waste emission and rapidly deteriorating environmental pollution. The recycling of waste in this period was not only a significant supplement to the supply of raw materials but also a vital method to reduce the emissions of pollutants.

From the perspective of environmental protection, this stage can be called the formation stage of China's environmental protection strategy.

In order to meet the increasing demand for environmental protection, in 1982, China incorporated the State Construction Commission, State Administration of Urban Construction, National Administration of Surveying and Mapping, and the Leading Team for Environmental Protection under the State Council to establish the Ministry of Urban and Rural Development and Environmental Protection, one of the internal departments of which was the Environmental Protection Bureau. Two years later, in 1984, the Environmental Protection Bureau was reshuffled into National Environmental Protection Agency, which was still under the leadership of the Ministry of Urban and Rural Development and Environmental Protection. In 1988, the environmental protection responsibilities were severed from the Ministry of Urban and Rural Development and Environmental Protection to the newly founded National Environmental Protection Agency (a sub-ministerial level agency), which was upgraded to State Environmental Protection Administration (Ministerial level), as an integral department of the State Council in 1998.

In this period, environmental protection was gradually strengthened. The comprehensive utilization of resources also developed into an important means of environmental end treatment in addition to its function in addressing insufficient supply of resources.

On December 26, 1989, China promulgated the first *Environmental Protection Law*, of which Article 25 stipulates: "For technological transformation of newly-built industrial enterprises and existing industrial enterprises, the facilities and processes that effect a high rate of the utilization of resources and a low rate of the discharge of pollutants shall be used, along with economical and reasonable technologies for the comprehensive utilization of waste materials and the treatment of pollutants." In 1995, the state formally enacted the *Law of the People's Republic of China on the Prevention and Control of Environmental Pollution by Solid Waste*, of which Article 3 provides: "The state shall, in preventing and controlling environmental pollution by solid waste, implement the principles of reducing the discharge of solid waste, fully and rationally utilizing solid waste, and making it hazardless through treatment." Article 4 stipulates: "The state encourages and supports the comprehensive use of resources, full recovery and rational utilization of solid waste, and adopts economic and technical policies and measures that facilitate the comprehensive use of solid waste."

The promulgation and implementation of the above two laws make the comprehensive utilization of wastes one of the major means of pollution prevention and environmental protection. Although the idea of circular economy was not put forward directly, the comprehensive use of resources and recycling of wastes, both of which are important contents of circular economy, were advocated and promoted by law.

3.2. The dual orientation of resources and environment: from classical stage to modern stage (1998-2004)

3.2.1. Inspiration from "Midnight Action for Pollution Control in Huaihe River"

Originated from the first major pollution accident of the Huaihe River in 1989 (when the waterworks were shut down and the lives of millions of people were seriously threatened), the "Midnight Action for Pollution Control in Huaihe River" began at the "on-the-spot" meeting for the inspection of pollution control and law enforcement in the Huaihe River Basin held by the Environmental Protection Commission of the State Council in May 1994, and ended at the midnight of the last day of the twentieth century when the Huaihe River was supposed to turn into a clean river as a myth. It lasted for several years, and ended up in vain, leading to the 2015 nationwide discussion on the investment of RMB60 billion in fruitless into the decade-long pollution control in Huaihe River.

The result of "Midnight Action for Pollution Control in Huaihe River" was not satisfactory, and the great revelation to the environmental protection management authorities is that it is not only impossible to establish a long-term mechanism but also costly to regulate the environment by means of administrative coercive measures such as closure and transfer.

The development of economy and the reduction of pollutants in their generation and emission can only be possible by addressing the problems from the source, promoting clean production and efficient utilization of resources, and recycling and re-using wastes from production and lives. Therefore, efficient and safe recycling of resources is highly compatible with environmental protection and people's pursuit of a better life. Around 2000, this conclusive understanding of China's environmental authorities turned traditional means of environmental protection to a circular economy development model which guided enterprises to recycle resources.

3.2.2. The stage transferring from end treatment to cleaner production and circular economy (1998–2002)

In view of the low efficiency and high cost of end treatment, domestic scholars introduced the concept of clean production in the 1990s, as well as the concepts of circular economy and circular society from Germany and Japan around 2000. On June 29, 2002, the 28th Meeting of the Standing Committee of the Ninth National People's Congress formally promulgated the *Law of the People's Republic of China on Promoting Clean Production*, which provided: "Clean production refers to reducing pollution from the source, raising the efficiency of utilizing the resources, reducing or avoiding the production and emission of pollutants in the process of production, services and using products by means of incessantly improving designs, using clean energy and raw materials, adopting advanced techniques, technologies and equipment, improving management, making comprehensive utilizations, and other measures so as to alleviate or eliminate the harm done to the health of the human being and the environment." This is the first special law in China with pollution prevention as its main content. While promoting cleaner production, the State Environmental Protection Administration started to push forward circular economy by organizing various training courses and seminars, popularizing the concept and knowledge of circular economy, with pilot projects to be done in some enterprises and industrial parks.

Different from Western countries that developed circular economy which mainly focused on waste disposal and comprehensive utilization, China, in view of the fact that its environmental pollution originated from industrial production at that time, started from resource allocation and enterprise layout in key industrial fields, and constructed, according to the principle of circular economy, a material recycling consortium based on industrial chain. Then, it launched an industry system based on resource recycling in line with the principle of reduction, reuse and recycling. With the building of specialized renewable resources industrial parks for domestic wastes, the country witnessed an upsurge in the development of circular economy nationwide.

3.2.3. The stage of preliminary formation of modern circular economy strategy with dual orientation of resources and environment (2002–2004)

In 2001, Chinese economy gradually recovered from Asian financial crisis and entered a new round of high-speed growth since 2002, which featured the accelerated urbanization and the upgrading of consumption structure, with demand in investment and consumption based on the heavy chemical industry. Therefore, along with the rapid growth of economic aggregate came the sharp increase in resource consumption and wastes produced, as well as the concurrent shortage of resources and the aggravation of environmental pollution. The price of renewable resources from wastes as raw materials kept rising too, with the environmental bearing capacity being approached to its upper limits. This historical background coincided with the time when the concept of circular economy was introduced and put into practice. As a result, circular economy could not only increase the supply of renewable resources but also could prevent pollution emission at the very source. The decision-makers were more determined to vigorously develop circular economy to solve the increasingly sharp contradictions between economic growth, resource shortage and environmental pollution. On October 16, 2002, President Jiang Zemin, then General Secretary of the CPC Central Committee, pointed out in his speech at the second meeting of the Global Environment Facility: "Only by taking the path of circular economy based on the most effective use of resources and environmental protection can sustainable development be achieved." In 2003, President Hu Jintao, then General Secretary of the CPC Central Committee, stressed at the National Symposium on Population, Resources and Environment: "We should speed up the transformation of economic growth mode and apply the concept of circular economy into the economic development of regions, the construction of urban and rural areas, and in product manufacturing, so as to make the most efficient use of resources; we should minimize waste discharge, gradually make the ecological cycle into a virtuous circle, and strive to build environmental protection model cities, ecological demonstration zones, ecological provinces." The *Decision of the CPC Central Committee on the Enhancement of the Party's Governance Capability* adopted at the Fourth Plenary Session of the 16th CPC Central Committee formally stated to make "resources conservation and environmental protection, the vigorous development of the circular economy and the building of a conservation-conscious society" as a major content of adhering to Scientific Outlook on Development and improving the Party's capacity to manage the socialist market economy.

Under the vigorous promotion of the national environmental protection authorities at that time, it has become a national consensus to vigorously develop circular economy. However, the function of the environmental protection department decided that it could not solve a series of problems in coordinating economic policies and interest relations encountered in the development of the circular economy, so by the second half of 2004, the State Council decided to have the development and management functions over circular economy transferred from the State Environmental Protection Administration to the National Development and Reform Commission. This transformation of the management system and mechanism has brought the development of circular economy into the national core strategy and opened a new stage of the development of circular economy in China.

3.3. The stage of rapid rise of circular economy (2005–Present)

3.3.1. Circular economy has developed from being a method of waste management to the mode of economic development

In September 2004, the National Development and Reform Commission held the first National Working Conference on Circular Economy and proposed to guide the preparation of the Eleventh Five-Year Plan with the concept of Circular Economy. In March 2005, then Premier Wen Jiabao put forward in the *Report on Government Work* that the development of circular economy should be taken as a measure to implement the Scientific Outlook on Development, transform economic development mode and relieve the pressure on resources and the environment. In July 2005, the State Council formally issued the *Opinions on Accelerating the Development of Circular Economy*, and laid down principles and guidelines regarding the objectives, key areas and management measures for developing Circular Economy in China. Since then, circular economy has been established as an integral part of the new economic development model and a central link of the major development strategies in China.

3.3.2. The Eleventh Five-Year Plan included circular economy in the national strategy

The *Outline of Eleventh Five-Year Plan for the National Economic and Social Development of the People's Republic of China* approved by the National People's Congress in March 2006, specially established a chapter on the development of circular economy, which constituted an overall deployment for the development of circular economy and resource-conserving society during the Eleventh Five-Year

Plan period. The *Outline* pointed out that vigorously developing the circular economy and speeding up the construction of a conservation-oriented society was not only an important measure to implement the Scientific Outlook on Development, which is people-oriented and of comprehensive coordination and sustainable development, but also the inevitable choice to realize the transformation of economic growth mode, fundamentally mitigate the resource constraints, reduce environmental pressure, promote the rapid and sound development of national economy, and achieve the goals of building a well-off society in an all-round way and of sustainable development. According to the *Outline of the Eleventh Five-Year Plan*, the National Development and Reform Commission was responsible for comprehensively popularizing the concept of circular economy, organizing pilot demonstration work throughout the country, and shifting the development of circular economy from concept and theory research to large-scale practice.

3.3.3. Circular economy has been promoted in accordance with law

In order to make the development of circular economy rule-based and law-based, and to build a long-term mechanism suitable for it, on August 29, 2008, the fourth meeting of the Eleventh NPC Standing Committee voted on and approved the *Circular Economy Promotion Law of the People's Republic of China* (hereinafter referred to as the *Circular Economy Promotion Law*) which would be formally enacted as of January 1, 2009. This is the third national legislation for circular economy in the world and the first of its kind among developing countries. The *Circular Economy Promotion Law* clearly stated that it was imperative to adhere to the basic principle of "reducing, reusing and recycling," and to establish the major systematic arrangements including the preparation of circular economy development plan, total volume control, evaluation indicator, extension of producer responsibility, key supervision and administration system, energy administration and standards of circular economy and so on, providing a legal and institutional guarantee for accelerating the development of circular economy. The *Circular Economy Promotion Law* promoted the development of circular economy to the strategic height of national economic and social development.

3.3.4. A special national plan has been formulated for the development of circular economy

On December 12, 2012, the Executive Meeting of the State Council specially studied and deployed the development of circular economy. The meeting pointed out that the development of circular economy is an important strategic task for China's economic and social development, and is an important and basic way to promote the building of ecological conservation and to realize sustainable development. The meeting discussed and adopted the *Circular Economy Development*

Strategy and Near-Term Action Plan, which was made available to public in January 2013. The action plan put forward four tasks of developing circular economy during the Twelfth Five-Year Plan period, namely, constructing circular industrial system, circular agricultural system, circular service industry system and carrying out demonstration work of circular economy. The plan also proposed the improvement of policies regarding tax, finance, industries, investment, price and fees, the enhancement of regulatory standards, the establishment of statistical evaluation system, the strengthening of supervision and management, the active international exchange and cooperation, and the comprehensive promotion of guarantee mechanism for circular economy.

3.3.5. An action plan has been formulated for circulating development

According to the guiding principles of the Fifth Plenary Session of the Eighteenth CPC Central Committee, China's NDRC formulated an action plan for circulating development during the Thirteenth Five-Year Plan period. The plan represented an effort to implement the new development philosophy emphasizing innovative, coordinated, green and open development for all, promote the transformation of the growth model, improve the quality and efficiency of development, take the lead in forming green mode of production and life, and facilitate the green transformation of the economy.

The action plan highlighted China's new mission, targets and paths for the development of circular economy in the new era, which could be summarized as follows. First, it facilitated the improvement of the circular economy at a higher level so as to build a circular industrial system and beef up institutional support. Second, it deemed circular economy as a crucial component to bring about new growth drivers and enabled sustained development of the circular economy by integrating new characteristics of economic growth in the new era. Third, it unveiled ten major actions for developing circular economy.

The plan displayed the role of circular economy in advancing the supply-side structural reform and improving the development quality, emphasized the combination of stronger institutional support and a strengthened role of market mechanisms, made clear the responsibilities and duties shared by governments, enterprises and individuals in the circular economy, and established a long-term mechanism to boost its development based on incentives and restrictions. In this plan, efforts were made to introduce a system for extended producer responsibility, establish a system for the promotion and use of renewable products and materials, improve the system on the limited use of disposable consumer goods, deepen the evaluation system for circular economy, strengthen the standards and authentication systems regarding circular economy, and promote the green credit management system.

The plan proposed ten major special actions to lead the development of circular economy during the Thirteenth Five-Year Plan period.

1. Action plan for circular transformation of industrial parks: By 2020, the state will mainly support the circular transformation of 100 industrial parks, and push 75% of national industrial parks and 50% of provincial industrial parks to carry out circular transformation.
2. Action plan for construction of circular economy demonstration zones for industrial and agricultural compound: To construct twenty circular economy demonstration zones for industrial and agricultural compounds.
3. Action plan for construction of industrial bases for resource recycling: Urban resource recycling industrial bases will be built in 100 cities at the prefecture level and above.
4. Action plan for construction of industrial bases for the comprehensive utilization of industrial wastes: To build fifty industrial bases for the comprehensive utilization of industrial wastes, and carry out major demonstration projects for the comprehensive utilization of industrial wastes.
5. "Internet +" resource recycling action: To formulate and release the *"Internet +" Resource Recycling Action Plan*, support the recycling industry to build recycling networks that integrate online and offline recycling, and promote the new mode of "Internet + recycling."
6. Action plan for coordinated development of circular economy in the Beijing-Tianjin-Hebei region: To make overall planning for the facilities that are used for the utilization and non-hazardous disposal of renewable resources, industrial solid wastes and domestic wastes in the Beijing-Tianjin-Hebei region, and build a batch of cross-regional demonstration projects on the comprehensive utilization of resources and coordinated development.
7. Action plan for the promotion of regenerated products and remanufactured products: To build about thirty promotion platforms and demonstration and application bases for regenerated products and remanufactured products.
8. Action plan for innovation in resource recycling technologies: To improve resource utilization efficiency and resource recycling level, carry out R&D for equipment technology of major generality or bottleneck technology in the circular economy, strengthen the demonstration of the integrated model of circular development in different regions.

9. Action plan for the promotion of typical experience and modes of circular economy: To summarize typical experience of pilot or demonstration projects of circular economy, as well as the typical cases and modes of circular development in key industries, and promote them in the whole society based on actual work practice.
10. Action plan for system innovation of circular economy: To carry out system innovation of circular economy in selected regions and industries, construct pilot zones, explore and form the core systems of circular economy and gradually promote them throughout the country.

4. MAJOR ACHIEVEMENTS

4.1. China has led the world with its widely recognized concept of circular development

After nearly twenty years of theoretical research and practice, China has formed a thinking system that includes developing circular economy vigorously, realizing green circular low-carbon development and building ecological civilization and a beautiful China. The country's achievement based on this thinking system has blazed a trail successfully for developing countries to follow in their pursuit of clean production and to work together with developed countries in tackling climate change. The road of China has been recognized all over the world.

In just forty years, through the continuous top-down and bottom-up efforts and under the direct leadership and promotion of the decision-makers at central government level, China has embarked on a development road with its own characteristics, including raising public awareness of environmental protection by disseminating the knowledge of circular economy, establishing and improving laws and institutions, comprehensively encouraging technological innovation, advancing pilot and demonstration projects, highlighting the top players' exemplary roles and promoting experience nationwide. Through compressed industrialization, China has completed transformations constantly in its rapid growth, meeting the needs of economic growth, energy conservation and environmental protection, thus achieving win-win results with major contributions to the formation of the nation's confidence in its road, theories, institutions and cultures.

Since 2005, with the approval of the State Council, the National Development and Reform Commission and other relevant departments have organized various pilot and demonstration projects on circular economy, with the participation of nearly 1,000 institutions and organizations. Localities have also carried out corresponding pilot projects and demonstrations, involving a wide range of administrative authorities at different levels, all key industries, fields and enterprises, with the emergence of a number of exemplary organizations, models and typical experience of circular economy. The National Development and Reform Commission has summarized sixty cases regarding the replicable and promotable models of circular economy as well as the experience from eight categories of pilot and demonstration projects. Led by pilot and demonstration projects, China has preliminarily shaped a development mode of circular economy, which is characterized by reduction at source, process control and the end regeneration and utilization and covers such sectors as industry, agriculture and service, such links as production, circulation and consumption, and various levels of enterprises, industries, industrial parks and society.

In recent year, China has increased its intellectual assistance to developing countries in their industrialization, providing special training on circular economy to thousands of civil servants at and above mid-level in developing countries. After the financial crisis broke out in 2008, the developed world highly recognized China's experience and gradually accepted the concept of circular economy in the restructuring of so-called "reindustrialization." For example, the World Economic Forum in Davos, a believer of circular economy, has organized special forums on circular economy successively in the past few years, disseminating the concept of circular economy by publishing research reports continuously. McKinsey & Co., a prominent international consultancy company, deems circular economy to be an opportunity for industrial innovation, suggesting that Europe accelerate its transition to circular economy. In 2015, the European Commission submitted to the European Parliament and the European Council *Closing the loop - An EU Action Plan for the Circular Economy*, which is equal to an indirect recognition of China's concept of circular economy. In 2016, *Nature*, an internationally renowned scientific journal, introduced the circular economy in the cover article, highlighting China's experience in developing circular economy. It recognized that the shift from linear economy to circular economy was the only solution to the resource security problem facing the world, and China's circular economic strategy was an important step to bridge the gulf between global economic development and ecological protection.

The theory of circular economy has become the thinking system which China imported, digested, absorbed, re-innovated and exported.

4.2. China has formed an array of advanced models through pilot and demonstration projects

Circular economy is an economic development model that fully integrates the efficient circular utilization of resources with environmental protection. Pouring huge efforts on various kinds of pilot and demonstration work on circular economy, China has formed a series of mature and advanced models.

4.2.1. Development model of circular economy for enterprises

The development mode of circular economy for enterprises refers to a single or key enterprise' circular productive process which starts from ecological design and reduction at the source, to the improvement of process control and the end-recycling, and then to the material circulation between the processes and production units, which helps the production chain to be extended, and then to the construction of a network for the circular utilization of material and cascade utilization of energy. All of these contribute to the improvement of energy utilization efficiency, the reduction in the input of material and energy as well as in the emission of waste and toxic substances. Therefore, it is a production mode that can raise the competitiveness of an enterprise.

For example, the steel industry, which is traditionally considered to yield the most pollutants, has greatly reduced its emission intensity of pollution gases such as SO_2 and NO_x by developing the long-process-based steel enterprise consortium and technically realizing desulfurization and denitration in raw material inlet sintering and coking processes. Coke oven gas, blast furnace gas and converter gas are all recovered for power generation or used as chemical raw materials. Blast furnace water granulated slag and steelmaking slag are both recycled and efficiently utilized. All water resources are classified for reuse. Sludge is also used as raw material. Residual heat and residual pressure, in addition to their cascade utilization, along with wastes, have become the main profit source of the iron and steel consortium. Thanks to the development of circular economy and technological innovation, the comprehensive energy consumption of one ton of steel in China has been reduced from 2.35 tons of standard coal in 1978 to about 0.6 tons now, while the new water consumption for the production of one ton of steel has been reduced from scores of tons to about 3 tons. The smokestacks of iron and steel industry have changed from emitting black smoke to white smoke (steam). Steel plants have also evolved from producing solely steel ingots in the past to a consortium of steel "plates, pipes, bars, wires," power generation, building materials and other circular economic products.

4.2.2. Development mode of circular economy in industrial park

Industrial agglomeration in the industrial parks has become the basic pattern of productivity distribution in China's industrialization. There are thousands of industrial parks at the provincial level and above, and tens of thousands of them at the county level. But, along with such development came a saying that "industrial agglomeration is pollution agglomeration." In order to solve the problem of pollution emission agglomeration brought about by industrial agglomeration, the state has promoted "recycling-oriented reconstruction of existing stock in industrial parks and recycling-oriented construction of new increments" as the focus of developing circular economy since 2005.

What is the core of circular transformation and construction of industrial parks? The basis is to impromve the functional, industrial and enterprise layouts of the industrial parks and to extend the industrial chain. Then, material circulation and energy cascade network can be constructed within and among enterprises so that they can establish the relationship of material metabolism and symbiosis, share the environmental infrastructure, build platforms for the exchange and management of wastes and energy, promote the reduction, reuse and recycling of the raw materials and wastes, as well as the recycling and harmless treatment of hazardous wastes. In this way, the consumption of material, energy and water in the industrial parks can be minimized, along with better environmental pollution control and emission reduction, which improves the resource output rate and the comprehensive competitiveness of industrial parks. After ten years of promotion, the development of circular economy in industrial parks has achieved remarkable results, and the problem of pollution agglomeration has been greatly alleviated.

4.2.3. Cross-industry circular economy development model linking agriculture, industry and services

The traditional agricultural circular economy is mainly about the material circulation formed by the simple conversion of agricultural organic wastes like organic fertilizer and biogas from the planting industry, animal husbandry industry, and forest and fruit industry. Typical examples include returning straw back to field, aerobic composting of animal dung, "pig-raising, methane-generating and fruit-growing" system, "pig-raising, methane-generating and vegetable-growing" system and so on. However, economic value is not high under the system of household contract responsibility system due to the small farm scale and the low value of wastes recycled. Since 2010, however, guided by national demonstration projects, a number of circular economic consortiums featuring integrated development have been formed to cover sectors like agricultural cultivation, breeding, processing of agricultural products, production of organic fertilizers, utilization of biomass energy sources, food industry, construction materials manufacturing, tourism, catering services and health care. In addition, internet technologies, such as e-commerce and big data, have also been widely used in the compound circular economy system of agriculture, industry and service, which may be a major trend in the transformation of rural and agricultural development in the future.

4.2.4. Professional recycling mode of social waste resources

As living standards go up, various waste resources, mainly household wastes, have become an important source of environmental pollution as their quantity keeps growing. These wastes include used electrical and electronic products, used automobiles, waste plastics, waste rubber, waste paper, waste textiles, kitchen wastes, packaging waste, waste glass, and the wastes generated from industries and service sectors, like waste electronic and electrical equipment, waste metal, sludge from sewage treatment plant, medical waste and so forth. In the past, these wastes were mainly landfilled. But with the development of urbanization, cities were increasingly under the siege of wastes, leading to the improvement of residents' environmental consciousness and increasing NIMBY effect. Pressure on recycling and safe disposal wastes aggravates.

Since 2005, by learning the experience of developed countries, China has explored the use of "Internet +" among other technologies to establish a recycling system for waste resources, specialized urban mineral bases, and specialized bases for safe disposal and recycling of urban waste resources. As practice has proven, they are effective models for the treatment and recycling of low-value urban wastes.

4.2.5. Remanufacturing mode

The remanufacturing mode, as its name suggests, is a kind of recycling mode for waste resources, according to which a damaged product or equipment will be repaired in its working face or in parts so that it can resume its original functions or service life on the basis of original manufacturing. The remanufactured parts or products are as good as the original parts or products in terms of performance or quality. Let us take the engines remanufactured by Sinotruk Ji'nan Fuqiang Power Co., Ltd. as an example. Their remanufactured engines have a quality and function not lower than or even higher than that of the new engine. Compared with manufacturing the new engines, the remanufacturing of engines takes up only 50% in cost, only 60% in energy consumption and 30% in material consumption. Thus, they have very high economic, ecological and environmental benefits. At present, the remanufacturing industry has become a high-end industry of circular economy, with a layout all over the country. Remanufactured products have also been extended to include mining machinery, metallurgical equipment, petrochemical equipment, transportation equipment, mechanical and electrical equipment, among many other fields.

4.3. Significant benefits have been achieved in terms of resources and environment contributing to addressing climate change[4]

The development of circular economy has greatly improved the utilization efficiency of resources and reduced the amount of newly explored raw resources, environmental pollution and greenhouse gas emissions.

4.3.1. Effect of resource saving is notable

According to incomplete statistics in 2015, the total amount of industrial solid wastes reused in the whole year exceeded 3.1 billion tons, an increase of nearly 2.4 billion tons over 2005. From 2005 to 2014, a total of 790 million tons of scrap steel and 80.85 million tons of recycled non-ferrous metals, including recycled copper, recycled aluminum, recycled lead and recycled zinc, were utilized. In 2016, China comprehensively utilized more than 2 billion tons of industrial wastes, including 432 million tons of fly ash, 477 million tons of coal gangue, 353 million tons of metallurgical slag and 18.08 million tons of waste plastic. The country used 800 million tons of straw and other agricultural and forestry wastes and 2.1 billion tons of livestock manure. It is estimated that the total amount of energy recovered from wastes could be converted to more than 400 million tons of standard coal.

4.3.2. Environmental benefits are outstanding

In 2016, the total amount of solid waste resources reused was estimated to exceed 6 billion tons, with the reused reclaimed water exceeding 9 billion tons, and various flue gas and dust recovered and utilized through desulfuration and denitration exceeded 50 million tons, which played a pivotal role in improving the ecological environment. What is worth lauding, in particular, is that the recycling of a large amount of resource supply means less need to explore new resources, hence reducing the damage to the ecological environment at the very source.

4.3.3. Great contribution to the reduction of greenhouse gas emissions

A preliminary calculation of the greenhouse gas emission reduced by China's circular economy in 2014 shows that the comprehensive recycling and utilization of mineral resources, industrial wastes, agricultural wastes, renewable resources and garbage could be converted to as high as 1.3 billion tons of CO_2 emission reduced, an equivalent to 12% of the national CO_2 emissions in that year, or 16% in the decline in the intensity of greenhouse gas emissions per unit of GDP. Among them, the mineral resources reused dropped the greenhouse gas emission by 29 million tons, industrial wastes by 1.1 billion tons, agricultural wastes by 39 million tons, renewable resources by 38 million tons and garbage by 78 million tons. Meanwhile, in 2014, the emission of five greenhouse gases reduced through resource recycling reached 170 million tons of CO_2 equivalent. It is safe to say that the development of circular economy is one of the major ways to address the problems of greenhouse gas and climate change. For example, the typical long-process steel circular economy can reduce greenhouse gas emissions per unit of steel by more than 70% compared to the previous non-circular economic mode.

Using metallurgical slag to produce cement can reduce greenhouse gas emissions per unit of cement by over 80%.

5. EXISTING PROBLEMS

The development of circular economy involves multiple aspects, including but are not limited to the changes in industrial organization forms, innovation in resource allocation modes, formation mechanism of resource price system, environmental protection standards, technical structures, management systems, waste recovery systems and operation mechanisms, the interest relationship between waste producers and recyclers, the coordination of national policies and so on. However, it takes time for all stakeholders to adjust themselves to the needs of the development of circular economy. Therefore, circular economy is faced with some problems.

5.1. The development of circular economy needs improvement in the long-term mechanism and innovation in the institution

Although the state has promulgated the *Circular Economy Promotion Law* and the provinces, autonomous regions and municipalities have formulated circular economy promotion regulations and policies suitable for themselves, there are still some institutional obstacles to the development of circular economy.

First of all, the *Circular Economy Promotion Law* is not a mandatory entity law, and the specific subordinate entity laws and regulations systems are far from being perfect. For example, there is a lack of special laws on the extension of producer responsibility and on the recycling of bulk wastes. The government supports the development of circular economy with financial subsidy and tax preferential policies, which, however, are still temporary and responsive without an established corresponding standardized system.

Second, the taxation system needs to be amended. The low usage fee of resources in China leads to the low price of initial resources, which will compromise the comparative economic benefits of waste resources recycling.

Finally, the anti-driving mechanism of environmental end treatment is far from perfection. Since 2018, China has started to levy environmental protection tax, putting an end to the history of free discharge of pollution by enterprises that meet environmental standards. However, waste discharge remains free of charge for residents and non-business units, who pay currently only the garbage transportation fee instead of the garbage disposal fee. The institutional arrangements for the free discharge of waste from non-business units result in the fact that the subsidies for the cost of waste disposal and recycling are borne solely by the government, while such expenditure by the government is not legally bound. This leads to a lack of reliable financial support for the development of circular economy.

5.2. The cost of developing circular economy increases, leading to a decrease in the economic benefits of enterprises

China's economic development has entered a "new normal," which is characterized by the late stage of industrialization, the disappearance of the demographic dividend, the slowdown in the growth of previously key industries like heavy industry and real estate, ever harder resource and environment constraints, and a downward trend of economic growth rate. Meanwhile, international commodities, represented by oil and mineral products, have been oversupplied with prices continuing to fall. Against this backdrop, the cost of labor increases along with that of environmental protection when the market resource price is constantly on the decline, leading to the drop in the price of renewable resources and products. The profits of enterprises engaged in recycling renewable resources are engulfed by rising costs and falling benefits, leading to a graver situation. All these contribute to the serious restriction of the development of circular economy. For example, the market price of scrap steel, recycled plastics and building materials was reduced by nearly 50% compared to that in 2007, which caused the economic benefits of the enterprises involved in the circular economy to fall sharply, with some of them suffering serious loss of money.

5.3. Environmental rules are increasingly strict and market competition intensified

As China intensifies its efforts on the building of ecological civilization, the requirements of environmental protection are also becoming more rigorous, which has exerted a certain influence on the development of circular economy.

First, the increasingly strict environmental regulations lead to the growing intensity of environmental investment in waste resources recycling, as a result of which, the cost of converting wastes into renewable resources rises, seriously eroding and squeezing the profit of recycled products. The business environment for enterprises engaged in the recycling and comprehensive utilization of resources is deteriorated. In 2017, the State Council issued a notice prohibiting the entry of foreign wastes, blocking the supply channel of waste resources from overseas, which reduces waste resources from the source and intensifies competition in domestic market for renewable resource processing and manufacturing.

Their economic benefits have also declined. The adjustment might take quite a long time.

Second, market competition becomes fiercer, leading to cumulative risks in the industrial chain of circular economy. The closer the circular economic complex, or the longer the circular economic industrial chain, the greater the market risk, for the failure of any enterprise in any link of the circular economic system or in the circular economic industrial chain due to the shrinkage of the product market might cause the collapse of the whole system and the complete industrial chain, which will make the circular economic system difficult to operate.

5.4. Technical bottlenecks restrict the development of circular economy

As a new technical and economic paradigm, circular economy needs to be, in the first place, technically feasible, economically reasonable and profitable. The economic return of resource recycling often depends on technical efficiency, which means that technology must make the recycling of resources and waste more efficient than the use of new resources, at least in the case of preferential policies determined by the system. Only in this way can circular economy be sustainable. At present, there are technical bottlenecks in the recycling of waste resources in China, which makes the cost of recycling too high, thus restricting the development of circular economy. For example, economic benefits are constrained by technical bottlenecks in such fields as recycling of used electronic products and used batteries, backfilling of mines with tailings, recycling of toxic and harmful non-ferrous metal slag, and high-value utilization of fly ash.

5.5. The implementation of policies is not in place, affecting the development of enterprises

As the development of circular economy is a matter concerning all aspects of the society, the short-term preferential policies for circular economy formulated by the government affect various departments of the government and stakeholders, with some policies increasing the cost and bringing down the performance of the departments under the current indicators system. Therefore, some policies are difficult to implement in practice, which greatly affects the enthusiasm of the enterprises in the circular economic industries. The preferential policies for circular economy are not legally binding to the state's finance, meaning that the implementation of these policies is very arbitrary. For instance, the state has rolled out a policy to expand the subsidies for waste electronic and electrical product from the previous "TV set, washing machine, refrigerator, air conditioner and computer" to fourteen kinds of electronic and electrical products. The policy, however, has not been implemented even if it has been released for more than a year. Even the subsidy for "TV set, washing machine, refrigerator, air conditioner and computer" is often paid belatedly or not paid at all, leading to the rise of financial cost of enterprises by a large margin. In 2015, the Ministry of Finance issued a preferential policy requesting "immediate refunding" of part of the value-added tax levied from enterprises in waste recycling business. Many provinces and cities, however, never implemented this policy.

The problem that the policies for circular economy are not well legally grounded, arbitrary, and often casual in implementation seriously limits the development of circular economy.

6. THE FUTURE DIRECTION OF DEVELOPMENT

6.1. Supporting ecological conservation

As early as 2005, Xi Jinping pointed out at the Working Conference on Circular Economy in Zhejiang Province that "we should deeply understand the strategic significance of developing circular economy, which is the concrete practice of implementing scientific outlook on development and constructing a harmonious society, the inevitable choice of changing growth mode and realizing sustainable development, and the urgent need to meet the challenge of economic globalization and enhance international competitiveness." Xi Jinping's thought on the development of circular economy is in line with the building of ecological civilization.

The report of the Nineteenth CPC National Congress clearly points out that as socialism with Chinese characteristics has entered a new era, what China now faces is the contradiction between unbalanced and inadequate development and the people's ever-growing needs for a better life.

Chapter IX of the Report of the Nineteenth CPC National Congress defines the strategy of "speeding up reform of the system for developing an ecological civilization and building a beautiful China." This strategic goal is to solve the imbalance between economic development and environment, and address the contradiction between the people's need for a better life and the imbalance of the supply of environmental benefits.

According to Xi Jinping's remarks on circular economy, the future development of circular economy will play an important supporting role in the building of ecological civilization in China. As stipulated in the task of building a beautiful China proposed by the Report of the Nineteenth CPC National Congress, the development of circular economy in the future should prioritize green development by establishing and improving the economic system of a green, circular and low-carbon development, promoting the comprehensive conservation and circular utilization of resources, implementing national water-saving action, reducing the consumption of energy and material, and realizing circulation link between production system and living system.

6.2. Establishing an institutional system compatible with the development of circular economy

According to the strategy of speeding up reform of the system for developing an ecological civilization and building a beautiful China in the Report of the Nineteenth CPC National Congress, the construction of an institutional framework and legal guarantee suitable for the development of circular economy and ecological conservation will serve as a major pillar for the development of circular economy in the future. Therefore, it will be an important direction for the future system reform to revise and improve the *Circular Economy Promotion Law* and formulate specific subordinate entity laws and regulations. Specifically, China will continue to improve and implement the environmental protection tax law, build and advance the system of extending producer responsibility, establish the preferential system of state fiscal revenue supporting the expansion of circular economy, reduce the arbitrariness in policy implementation, perfect special laws and regulations on recycling of bulk wastes, and establish a legally binding system for the participation by the whole people.

6.3. Promoting the successful development mode of circular economy in an all-round manner

China shall summarize, improve and promote the successful modes formed in the practice of developing circular economy in the past ten years before disseminating them in an all-round way in the future economic development. According to Xi Jinping's statement at the Working Conference on Circular Economy in Zhejiang Province, China shall embark on a way of scientific development by integrating the concept of circular economy into economic development, urban and rural construction and social production, and shall actively promote the development of circular economy and accelerate the transformation of growth mode.

In particular, in the planning of urban and rural construction and the construction of industrial parks, it is necessary to fully reflect the objective demands of resource recycling and environmental protection, and build a circular infrastructure system based on the circular transformation of the park stock and the incremental circulation. The state will also construct green industrial system and green agricultural system in accordance with the characteristic town of rural complex; advocate and popularize green consumption mode; vigorously develop green architectural system; vigorously promote the application of low-carbon technology; implement rehabilitation and ecological restoration projects in a scientific manner and so on.

6.4. Promoting innovation in the service mode of the circular economy

Circular economy involves every enterprise, public institution, social organization and citizen. Coordinated action by everyone will enable the effective management and utilization of waste resources. The market economic mechanism is ineffective in many links of waste recycling and disposal. While making up for the market failure through system innovation, it needs the support of a sound circular economic service system and improved models, the most important of which is the recycling system of waste resources.

The waste recycling system includes sound systems of information service, collection and classified transportation, resource trading and marketing service for circular economic products. Unlike the circulation service system of general commodity, waste exists in every household, every social organization, and the waste collection and classification involves every waste producer, whose cooperation is a must to make recycling services efficient. But this is a big and difficult problem. The pilot program of garbage sorting in China has been carried out for more than ten years but yielded little effects. For circular economy to achieve sustainable and efficient development in the future, we need to establish a perfect and efficient waste recycling system in the whole country by taking advantage of "Internet +" and other advanced technology systems and through technological innovation and mode innovation with the support of the government.

7. CONCLUSION

The development of circular economy in China has transformed from the recycling and utilization of waste resources to the mode of resource utilization, from resource orientation to environmental protection orientation, from demonstration through pilot projects to universally accepted economic development mode, from paying attention to economic benefits only to paying attention to environmental benefits and economic benefits at the same time. China's concept of circular economy has been widely recognized throughout the world. However, as Chinese economy goes through different stages, the development of circular economy is also under new tests. The first one is who pays for waste recycling. As the country's industrialization is coming to its end, the growth rate of demand for resource-based products decreases. Even the total consumption of some basic resources begins to decline. Although various kinds of industrial and domestic wastes continue to increase in quantity, they are declining in value as resources. The cost of recycling and reusing these wastes is increasing, but the benefit is decreasing. The environmental attributes of circular economy are becoming stronger, but the problem of who pays for recycling waste has not aroused wide attention or received enough positive feedback. The recycling of waste is facing the conceptual and practical tests of innovation in cost payment system. The second test is about the technological innovation in waste recycling and reuse as we will have to attend a comprehensive test on new technology, new mechanism, new model and capacity building. Further research needs to be done against the strategic background of the building and system innovation of ecological conservation.

Notes

1 *The Complete Works of Marx and Engels (Volume 25)*, People's Publishing Press, 1979, p. 95.
2 Pearce et al. (1996). *Green Economy Blueprint*, trans. He Xiaojun, Beijing: Normal University Press.
3 Wang Jun (2007). *Theory and Research Methods of Circular Economy*. Economic Daily Press.
4 Data in this section are retrieved from China Association of Circular Economy.

CHAPTER FIVE

Water Resources: Exploitation and Protection

XIA JUN AND ZUO QITING[*]

1. INTRODUCTION

Water is the source of life, the necessity of production and the basis of ecology. Without water, all living organisms, including human beings, would not survive; industry and agriculture would wither; and the virtuous cycle of ecological environment would stop. Water is an indispensable resource for human survival and development. Mankind has been dealing with water since its birth, and has accumulated in its age-old evolution rich experience and knowledge in the exploitation and utilization of water resources, which provided the driving force and support for further tapping into water potentials. Particularly, along with the development of economy and society in modern times, which enhanced human's productive force in transforming the mother nature, human has demonstrated

---

[*] Xia Jun, Ph.D., academician of Chinese Academy of Sciences, professor of Wuhan University, whose research field is hydrology and water resources, proposes the theory and method of hydrological nonlinear system identification, reveals the hydrological nonlinear mechanism of runoff formation and transformation, and develops the time-variant gain hydrological model and water system method. Zuo Qiting, PhD, professor of Hydrology and Water resources at Zhengzhou University, proposes quantitative research methods for sustainable use of water resources and harmony between man and water, establishes the discipline system of water science, the most stringent water resources management system theory, and develops the theory of adaptive utilization of water resources.

an ever-greater wish to modify the world, leading to an increasingly large scale of exploitation and utilization of water resources. However, water problems appeared as a result of such exploitation and utilization, which forced people to refrain from unrestrained demand for water and take protective measures. Therefore, human's "utilization" and "protection" of water resources are both related and contradictory, which means that we can only guarantee the sustainable use of water resources through protection in utilization and utilization with protection.

Since China adopted the reform and opening-up policy in 1978, great economic and social changes have taken place. In 1978, China had a population of 962.59 million, with a GDP of RMB367.87 billion and a per capita GDP of RMB385. In 2016, China had a population of 138.271 million, with a GDP of RMB74.41272 trillion and a per capita GDP of RMB53,980, which were 1.4 times, 202 times and 140 times that of 1978, respectively.[1] At the same time, the development and utilization of water resources have also changed greatly. In 1978, the water consumed by agriculture was 419.5 billion m^3, and that by industry was 52.3 billion m^3; in 2016, the figures changed into 376.8 billion m^3 and 130.8 billion m^3, respectively,[2] with the former reduced by 10 % and the latter increased by 1.5 times. During the forty years of reform and opening-up from 1978 to 2018, great changes have taken place in water resources' utilization and protection actions, concepts, theoretical researches and practice in China. Therefore, for the scientific understanding, rational utilization and effective protection of water resources, it is of great significance to summarize the experience and lessons we have learned about water resources utilization and protection in the forty years of reform and opening-up and to envision the future development as well as for the sustainable social and economic development.

There are not many literatures on the stages of water resources utilization, protection or water conservancy in China, but some related discussions do exist. For example, Xia Jun and others divided the formation stage of researches on and the science of water resources into the following stages: the stage of accumulating ancient knowledge of water resources (before the Westernization Movement in 1860), the embryonic stage of modern water resources researches (1860–1949) and the establishment stage of modern water resources science (after 1949).[3] Cao Xingrong defined the development and utilization of water resources into three stages: primary stage, basic equilibrium stage and water shortage stage.[4] In addition, some researchers analyzed the stages of water resources utilization in specific regions. Qu Yaoguang and others divided the development and utilization of water resources in the arid area of northwest China into three stages: the stage of surface water development and utilization, the stage of joint development and utilization of surface and underground water, and the stage of economical utilization of water resources.[5] Zhu Meiling split the water utilization in Habahe River basin into four stages: the stage of ecological and natural balance, the stage of imbalance, the stage of deterioration and the stage of restoration, and she judged that the fifth stage would be the benign development.[6] When it comes to the researches on the development stage of water conservancy in China, Wang Yahua divided the development of water conservancy into four stages: the period of large-scale water conservancy construction (1949 to 1977), the period of relatively stagnant water conservancy construction (1978 to 1987), the period of increasing contradiction in water conservancy development (1988 to 1997) and the period of water conservancy reform, development and transformation (1998 to 2010).[7] Zuo Qiting divided the development of water conservancy since 1949 into three stages: the engineering water conservation stage (water conservancy 1.0), the water resources conservancy stage (water conservancy 2.0) and ecological water conservancy stage (water conservancy 3.0), and proposed the intelligent water conservancy stage (water conservancy 4.0).[8] In addition, there are also special studies on the stages of water conservancy development during the country's reform and opening-up. For example, Chen Lei divided water conservancy during the course of China's reform and opening-up from 1978 to 2008 into three stages: the stage of difficult launching of water conservancy reform and development (1978–1987), the stage of deepening reform and development (1988–1997) and the stage of accelerated reform and development (1998–2008).[9]

Viewed from the present literature, there is a lack of summary and demarcation of stages of water resources utilization and protection in forty years of reform and opening-up. Therefore, based on the previous research results and by drawing on a large amount of literature, this chapter elaborated on the development course of water resources utilization and protection in China over the past forty years, and presents a demand-based prospective analysis on the coming stage of water resource utilization and protection in the new era.

2. THE COURSE OF DEVELOPMENT

2.1. Division of development stages and their main characteristics

According to the analysis of the trends in economic and social development, as well as that in water resources utilization and water environment changes since 1978, we can at least come to some conclusion about the evolution process and innate rules. First, the great strategic plan of reform and opening-up which was put forward in 1978 led to liberated ideas, opened markets, invigorated economy, a large-scale and rapid economic boom, and doubled economic aggregate with nature under expanding transformation and environmental engineering. The first ten years of reform and opening-up saw the focus laid on economy. To ensure a rapid growth, China launched a large-scale exploitation and utilization of natural resources. Then, in order to protect resources and improve the quality of economic growth, the country chose to sacrifice the growth rate for a stable economic development. Second, water resource utilization was not restrained or even rather imprudent at the beginning of the reform and opening-up, featured by sporadic casual development and even misuse, low utilization efficiency and the following pollution. Later, for the sake of water protection, China began to limit the total use of water resources, while at the same time improving the efficiency of utilization for the sustainable use of water resources. Third, when it comes to water environment change, the reform and opening-up brought about not just large-scale economic construction but also extensive use of water resources and random discharge of sewage, which led to sharp deterioration of water environment, seriously affecting water resources utilization and people's health and living. It has forced the public to behave in the development, protect water resources and water environment.

Based on the above development facts, we have divided the development course of water resources utilization and protection in China from 1978 to 2018 into three stages by analyzing the main representative events of different times. First, the development-focused stage (1978–1999): with water engineering and water resources development as main characteristics. Second, the comprehensive utilization stage (2000–2012): focusing on the comprehensive utilization of water resources with the goal of the harmony between human and water resources. Third, the protection-focused stage (2013–2018): use of water resources with the goal of protecting water ecology and building ecological civilization (see Table 5-1).

2.2. The development-focused stage centering on water engineering and water resources development (1978–1999)

When it was founded in 1949, the People's Republic of China was poor and backward and was faced, in particular, with huge difficulties in infrastructure construction, industrial and agricultural production caused by the economic blockade by capitalist countries. In order to restore production as soon as possible, the state concentrated its efforts on renovating and reinforcing river dykes, farmland water conservancy projects and large-scale water engineering projects, contributing greatly to the construction of new China. Water conservancy undertakings, including flood and drought control, farmland water conservancy, urban water supply and hydropower were thriving, which played an increasingly important role in the national economic and social development. But during the "cultural revolution," the whole country was in a chaotic situation with economic development stagnated and some projects destroyed. When people's basic livelihoods were not guaranteed, poverty and backwardness were the real picture of social

Table 5-1 Development Stages and Representative Events of Water Resources Utilization and Protection in China

Time	Stage and characteristics	Important events (representative events)
1978–1999	Focusing on water engineering and water resources development	• In 1978, the Third Plenary Session of the Eleventh Central Committee of the CPC proposed to launch reform and opening-up initiative, which centered on economy and initiated China's large-scale economic construction. • The first national survey and planning of water resources was completed in the 1980s. With a clear understanding of the water resources in mind, China began to plan the use of water resources, but the focus of the planning was still the water resources development. • In 1998, a rarely-seen flood occurred in Yangtze River, Nenjiang River and Songhua River, which exposed the weak link of water engineering—a factor obviously lagging behind the economic boom. • The rapid development of economy, compared with the inadequate investment in water engineering, meant that the main objective during this period of time was to develop and utilize water resources. The speed of water engineering lagged far behind that of economic growth in China.
2000–2012	The stage of comprehensive utilization with the goals of comprehensive usage of water resources and the harmony between human and water resources	• Around 2000, the concept of sustainable utilization of water resources began to be applied in the practice of water resources development and utilization. • In 2001, that concept of harmony between human and water resources was formally incorporated into the philosophy of modern water management. The theme of China Water Week in 2004 was "harmony between human and water resources." • In 2009 and 2010, China suffered from large-scale, repeated and severe floods and droughts, which shocked the whole nation. In 2011, the No. 1 Document of the CPC Central Committee issued the *Decision on Accelerating Water Conservation Reform and Development*. • In January 2012, the State Council issued the *Opinions on Implementing the Strictest Water Resources Management System*, and made comprehensive and specific arrangements for implementing the strictest water resources management system.

(continued)

Table 5-1 Continued

Time	Stage and characteristics	Important events (representative events)
2013–2018	The protection stage to protect water ecology and promote ecological conservation	• In January 2013, the Ministry of Water Resources issued the *Opinions on Accelerating the Construction of Water Ecological Civilization*, proposing to deployment for an accelerated construction of water ecological conservation. • In April 2015, the State Council issued the *Circular on Printing and Issuing the Action Plan for Prevention and Control of Water Pollution*, dealing a heavy blow to water pollution problems. • The report of the Nineteenth CPC National Congress held in 2017 proposed to "implement our fundamental national policy of conserving resources and protecting the environment, and cherish the environment as we cherish our own lives."

development at that time. The Third Plenary Session of the Eleventh CPC Central Committee held in 1978 put forward the strategy of reform and opening-up to shift the focus onto economic construction. In retrospect, this was a great choice made in the historical context of a relatively backward domestic economy and a very low living standard for the people.

In the first ten years of reform and opening-up, China focused mainly on reform at home and opening-up to the outside world. Economic development was an important measure to address the problem of unfulfilled basic needs of people for food, clothes, shelter and a better life. Economy grew fast, but the construction of water resources engineering projects was not as fast. Besides, the focus of work was mainly on the development and utilization of water resources, without as much consideration given to the protection of water resources.

When China entered the second decade (1988 to 1999) of reform and opening-up and intensive economic growth, it had accumulated much experience in large-scale development. Along with the accelerating economic expansion came the extensive exploitation and utilization of water resources, resulting in first, the excessive use of water and second, the discharge of sewage water exceeding the water environmental capacity, or specifically, water shortage, water pollution, soil erosion, ecological deterioration, floods, droughts and pollution accidents. Especially in 1998, the Yangtze River, the Nenjiang River and the Songhua River suffered from whole-basin floods rare in history, which caused serious loss of life and property.

During this period, in order to develop water resources, prevent flooding and support local economy, China completed some water projects nationwide, for example, the construction of irrigation and water conservancy facilities, the building of water supply facilities in urban and rural areas, the development of hydropower stations, water and soil conservation programs, flood walls, among others. All these projects were put into place rather quickly, such as Water Diversion Project from Luanhe River to Tianjin City (May 1982 to September 1983), Three Gorges Project of the Yangtze River (1994 to 2009), the Xiaolangdi Hydro Project on the Yellow River (1994 to 2001). In short, the investment in water conservancy engineering projects was barely adequate to cover areas other than the development and utilization of water resources, thus was lagged far behind economic development at that time.

During this period, China spared no efforts to provide science and technology support to water resources utilization and protection, especially under the guidance of concept of reform and opening-up, which emancipated the thought and expanded the research fields with academic flourishes. First, China carried out the first countrywide water resources assessment in the 1980s, officially kicking off a nation's thinking over the scientific utilization of water resources, which further laid a foundation for the research on the coordination between the utilization of water resources and socioeconomic development in China. Achievements made are as follows: the proposal of national outcome in 1985, the *Water Resources Assessment for China* published in 1987, two national water quality assessments completed in 1984 and 1996, *Water Resources Quality Assessment for China* published in 1996, and the industry standard "A Guide to Water Resources Assessment" published in 1999. Second, in terms of research on water resources system, Chen Shouyu introduced the fuzzy theory and proposed the fuzzy hydrology in 1987. Ding Jing acquainted us with the stochastic theory and published the "*Stochastic Hydrology*" in 1988. In 1985, Xia Jun applied the grey system theory in the water resources system and later published the *Grey System Hydrology*. Third, research on the impact of climate change on hydrology and water resources was initiated but did not attract enough attention, thus lagging behind that of foreign counterparts. Fourth, remote-sensing technology began to be applied in China's water conservancy in the early 1980s. Fifth, with the rapid development of simulation technology and its application in water resources, the research results of water resources optimal allocation theory, which has been widely and intensively applied since 1980s, are on the increase cumulatively. Sixth, the theory of sustainable utilization of water resources was studied with some preliminary results, but it was not intensively and extensively applied. Seventh, the concept and calculation method of water resources carrying capacity became the focus of water resources research at that time after being put forward and applied to water resources management and practice. Eighth, with the sharpening contradiction between supply and demand of water resources, the traditional "demand-based supply" principle was no longer appropriate, for simply relying on increasing water supply capacity was no more able to meet the social demand for water resources. Thus, the demand management mode of water resources was changed from passive supply management to active supply management.

On the whole, in the first twenty years of reform and opening-up, China at that time centered its efforts on economic growth, with the development of water resources, the increase of income and GDP as the major goals of the times. Insufficient consideration was given to water resources protection, first, because of China's large-scale economic construction and its urgent need to raise the living standards of people, and second, because of the low level of public understanding, which had not been transformed from the traditional concept that "water resources are inexhaustible." The concept of "sustainable utilization of water resources" discussed internationally at the end of the twentieth century was just introduced into China, and had not been well implemented in practical work, as a result of which, water resources protection did not gain the important position it deserved.

2.3. The stage of comprehensive utilization with the goals of comprehensive usage of water resources and the harmony between human and water resources (2000–2012)

Water disasters, water-related events as well as the changes in water situation that appeared at the end of the twentieth century brought new demand and impetus to the concepts, actions and work of water resources utilization and protection after entering the twenty-first century. First, around 2000, the internationally recognized concept of sustainable utilization of water resources began to guide and influence the practice of water resources utilization and protection in China. For example, national water resources planning, which began in 2002, adopted the guiding thinking of the sustainable utilization of water resources. In addition, the concept of comprehensive management of water resources, which was well received internationally, was also gaining prominence in China. Second, in the aftermath of the floods of Yangtze River, Nenjiang River and Songhua River in 1998, the government and academic community in China began amid woes to seriously analyze the water situation which the country was faced with as well as the countermeasures in their grasp. The central government made a major deployment for post-disaster rebuilding, the repair of rivers and lakes and the construction of water conservancy, which also changed, to certain extent, our traditional understanding. In particular, it has been gradually recognized that the unlimited use of water resources was not feasible, as water resources are not and cannot be regarded as "inexhaustible" resources and the development and utilization cannot exceed the carrying capacity of water resources. Third, with the improvement of people's living standard, we have basically solved the problem of food, clothing and shelter. Whereas the contradiction had previously been framed as a tradeoff between the needs of the people and China's "backward social production," it is now viewed as a tension between "unbalanced and inadequate development" and the "people's ever-growing needs for a better life," which is tantamount to the needs for better environment and the wish for a harmonious relationship between man and mother nature. Under these circumstances, human beings must protect the water effectively in their use of water. Against this backdrop, the idea of harmonious relationship between man and nature and of harmonious water control were put forward as a response to the new situation.

In 2001, the concept of harmony between human and water resources was formally incorporated into the guiding principles governing water control in China in the new century. The theme of China Water Week in 2004 was "harmony between human and water resources." Through the extensive publicity of the Water Week, more people developed an understanding of the harmonious relationship between human and water. Since 2001, China, while paying close attention to the comprehensive, rational and scientific utilization of water resources in the practice of water control, has adhered to the idea of harmony between human and water resources, which was defined as a goal of the modernized management of water resources. In academic circles, a number of research results on harmony between human and water resources have emerged since 2005, focusing on the concept and ideas of harmony between human and water resources and qualitative application in the early stage. With the development of quantitative research, the theory of harmony between human and water resources with quantitative research at its core has gradually formed.[10] It has been successfully applied to water resources' planning, allocation, dispatching and management, the strictest water resources management system, water environment capacity allocation and cross-boundary river water distribution and so on.

After the development in the first ten years of the twenty-first century, although China had laid special emphasis on the comprehensive utilization and protection of water resources, there were still an array of problems due to the limited national and local investment and insufficient efforts to protect water resources. From 2009 to 2010, China was stricken by large-scale heavy floods and droughts several times, including the severe nationwide droughts such as the once-in-thirty-year drought that spanned over autumn and winter in north China, the once-in-fifty-year autumn drought in the southern China, and once-in-ten-year early summer drought in Xizang, once-in-ten-year heavy drought in Hunan and Hubei provinces. On August 7, 2010, a severe debris flow disaster hit Zhouqu, Gansu province, resulting in serious casualties and property losses. So shocking were these floods and droughts that the central government made a scientific judgment that "we owe too much to water conservancy" and that "Vulnerable water conservancy facilities are still an obvious weaker area of national infrastructure." The No.1 document, the *Decision on Accelerating the Water Conservancy Reform and Development* released by the CPC Central Committee in 2011, stipulated that close attention be paid to the rational use of water resources and the harmonious coexistence between man and nature.

In January 2012, the State Council issued the *Opinions on Implementing the Strictest Water Resources Management System*, with overall and specific deployment and arrangements made for the implementation of the system. Its core content is the so-called "three red lines" and "four systems." In order to further curb water resource shortage and water environment pollution and from the perspective of "source management-process management-end management," the government proposed three red lines for water resources, namely, the red line for the control on exploitation and utilization, the red line for water efficiency and the red line for pollution control in water function areas. This is to set a "red line" for water resources management, provide an effective means for administration, and offer some support to the comprehensive utilization of water resources and the harmony between human and water resources.

During this period, in order to both utilize and protect water resources, an increasing number of water engineering projects were built, especially under the guidance of the concept of harmony between human and water resources. After the floods of the Yangtze River in 1998, the CPC Central Committee, the State Council and the people of the whole country all realized the utmost importance of the rational utilization and effective protection of water resources. Many measures were taken to harness large rivers, reinforce dilapidated reservoirs, build flood control works for key cities and toughen safety requirements in flood districts. As a result, the flood control system for great rivers in China was enhanced greatly but that for small- and medium-sized rivers was far from perfection, with the construction of farmland water conservancy under great pressure. The east and middle routes of the South-to-North Water Transfer Project were completed successively to solve the shortage of water resources and the deterioration of environment in the northern part of China, especially in the reaches of Yellow River, Huaihe River and Haihe River. In addition, in order to improve the conditions of water resources and curb the trend of water environment degradation, the Chinese government constructed some ecological water transfer projects, for example, the water transfer project from Bosten Lake to the lower reaches of Tarim River, which began since 2000; the emergency water transfer project from Nenjiang River to Zhalong wetlands; the emergency water transfer project from Yangtze River to Nansi Lake, which started in 2002; the emergency water transfer project from "Yuecheng Reservoir to Baiyangdian Lake" completed in 2004, and the "Yellow River to Baiyangdian Lake" started in 2006. Under the guidance of the concept of harmony between human and water resources, close attention has been paid to the comprehensive utilization of water resources, with equal attention, if not more, to the protection of water resources.

During this period, under the guidance of more science-based concepts of water control for the new century, China made great progress in water resources utilization and in the scientific and technological support for water resources protection. To begin with, in 2002, China made a national comprehensive planning for water resources, which bettered the techniques and methods of evaluation and planning of water resources and applied them nationwide. Next, the theory of sustainable utilization of water resources guided the planning and management of water resources in an all-round way. Third, researches on the theory and application of harmony between human and water resources began in 2005. In September 2006, the fourth Water Forum of China with the theme of harmony between human and water was successfully held in Zhengzhou, and published in a collection of papers titled "Theory and Practice of Harmony between Human and Water Resources." Since 2006, research on harmony between human and water has been growing in large numbers. Fourth, the pilot project of building water-saving society was launched in 2002, kicking off the theoretical and practical work related to water-conservative society. Fifth, with the transformation from engineering-focused water conservancy to resource-focused water conservancy, the traditional water resources management has also changed into the comprehensive management of water resources. The strictest water resources management system implemented in 2012 was a new mode of water resources management introduced in China. Last, research on the innate rules of water resources evolution under climate change and human activities has become a hot topic, where many research results have been produced.

In retrospect of the development in the first decade of the twenty-first century, we should realize more than ever the importance of the comprehensive utilization of water resources and should pursue the goal of harmony between human and water resources. We shall develop, utilize and protect water resources through the whole process from the control of total water use, the control of water efficiency to the control of total sewage discharge so that we can ensure the effective protection of water resources.

2.4. The protection stage to protect water ecology and promote ecological conservation (2013–2018)

On November 8, 2012, the report of the Eighteenth CPC National Congress fully elaborated the call for "vigorously promoting the building of ecological conservation." In January 2013, the Ministry of Water Resources issued the *Opinions on Accelerating the Construction of Water Ecological Civilization*, which laid down the deployment to speed up the construction of water ecological civilization, for which two batches of 105 pilot cities were selected throughout the country. In addition to the pilot work mentioned above, special emphasis was placed on the role of water ecology in the field of water engineering. It can be said that all water engineering planning, construction and management in later times have taken into account the ecological constraints and protection needs, which signify that China has entered the protection-focused stage, the goals of which were to protect water resources and build ecological conservation.

In April 2015, the State Council issued the *Circular on Printing and Issuing the Action Plan for Prevention and Control of Water Pollution*, which dealt a heavy blow to the problem of water pollution. With the improvement of water environment quality as the core, the action plan carried out rigorous supervision over sewage treatment, industrial wastewater treatment, overall control of pollutant discharge and many more, and started a strict accountability system so as to promote water pollution prevention and control, water ecology protection and water resources management. All these efforts have served as guarantee for the construction of a beautiful China with "ever blue sky, lush mountains and lucid water."

On December 5, 2016, the State Council issued the *Thirteenth Five-Year Plan for Ecological and Environmental Protection*, which put forward to implement the strictest environmental protection system with the improvement of environment quality as the core, and to wage three major wars on air, water and soil pollutions. All these moves aimed to strengthen ecological protection and restoration, strictly prevent and control ecological environment risks, and accelerate the modernization of national governance system and capacity in ecological and environmental fields.

On December 11, 2016, the CPC Central Committee and the General Office of the State Council issued the *Opinions on Full Implementation of the River Chief System across the Country*, requiring the nationwide establishment of the river chief system within two years. All provinces and municipalities have successively issued relevant documents to put the river chief system in place, open up the management mode of "one policy for one river," which achieved abundant results and effectively improved the ecological environment of rivers and lakes.

In 2017, the report of the Nineteenth CPC National Congress proposed to "implement our fundamental national policy of conserving resources and protecting the environment, and cherish the environment as we cherish our own lives," and once again stressed that "promoting ecological conservation is vital to sustain the Chinese nation's development." Protecting Ecology is the precondition of water resources utilization.

In the meantime, the State has carried out a series of engineering construction and system construction centering the building of water ecological conservation, water ecological protection and the strictest management of water resources. First, 105 pilot cities were selected nationwide for the building of water ecological civilization. Some provinces and cities rolled out their own pilot projects, with remarkable results achieved. Second, China implemented the strictest water resources management system throughout the country, with "three red lines" and "four systems" finalized. China finished the National Water Resources Monitoring Capacity Building Project (2012–2014), with three monitoring systems of water retrieval, water functional areas and the provincial cross-section of big rivers built finally. It is monitoring the amount of water drawn with drawing permit from more than 75% of the country's rivers and has realized on-line routine monitoring of water quality of the functional areas of more than 80% of the important rivers and lakes, as well as a basic on-line monitoring of the quality of important surface and drinking water sources, with full coverage of water quality monitoring of the provincial cross-sections of big rivers.

During this period, against the backdrop of protecting water ecology and promoting ecological conservation, China has made great progress in science and technology for water resources utilization and protection. First, in the process of building pilot cities of water ecological civilization, China has raised the theoretical systems and technical methods for the construction of water ecological civilization and applied them in practice. Second, China puts forward the strictest theory system, supportive system and institutional arrangements for water resources management, with the methods to determine the indicators of "three red lines" finally defined and implemented in practice. Third, China has accomplished tangible results in the further researches on the sustainable utilization of water resources, the carrying capacity of water resources, the harmony between human and water resources, the construction of water-saving society, the optimal allocation of water resources, the evaluation, demonstration, planning and administration of water resources.

On the whole, China has stepped up the publicity for water ecological protection since the idea of water ecological conservation was proposed in 2013. By issuing a series of policies and regulations to guarantee the construction of ecological civilization, China called on all people to take into consideration the influence and constraints of ecological environment in all human activities. The utilization of water resources during this period obviously displayed the characteristics of being "protection-focused."

3. ANALYSIS ON THE DEMAND FOR WATER RESOURCES UTILIZATION AND PROTECTION FOR DEVELOPMENT IN THE NEW ERA

On October 18, 2017, the Nineteenth CPC National Congress released a report (hereinafter referred to as the "the report to the Nineteenth CPC National

Congress "), which pointed out that "with decades of hard work, socialism with Chinese characteristics has crossed the threshold into a new era. This is a new historic juncture in China's development", when the "Thought on Socialism with Chinese Characteristics for a New Era" was proposed, which serves "a guide for all our Party members and all the Chinese people to take action as we strive to achieve national rejuvenation." In this new era, China faces many new demands in its development. The following is an analysis on the new demand for water resources utilization and protection in the new era from the perspective of several key words of social concern.

3.1. Reform and opening-up

The reform and opening-up policy, which was put forward at the Third Plenary Session of the Eleventh Central Committee of the Communist Party of China in December 1978, is a basic state policy of China. The great achievements of forty years have proved that the reform and opening-up accords with the characteristics of the times and the great trend of world development thus is the way to develop our nation and the powerful motive force for the progress of socialist cause, which has been affirmed by the report to the Nineteenth CPC National Congress, as it reads that "only with reform and opening can we develop China, develop socialism, and develop Marxism."

The Thought on Socialism with Chinese Characteristics for a New Era further defines "the overall goal of deepening reform in every field," on which we shall insist. Reform and opening-up is the general principle and policy of the socialist modernization drive, to which we must adhere for a long time. Reform and opening-up also bring new demands and opportunities to water resources' utilization and protection.

First, the water resources' utilization and protection need to be advanced in accordance with rapid development brought about by reform and opening-up, during the early stage of which, the change appeared from the countryside with the introduction of household contract responsibility system, which improved the efficiency of farming and water use, and reduced slightly the amount of water consumed for farming purposes. What came along was the reform of the state-owned enterprises, which endowed them with greater autonomy, activated the market, promoted the industrial expansion, and greatly enlarged the demand for water resources. But at the same time, the industrial reform led to growing water shortage and sewage drainage caused by excessive water consumption as well as the pollution of water environment triggered by the surge of discharge. All these have brought serious challenges to water resources utilization and protection since the beginning of the twenty-first century. In the next few years, the "construction of ecological civilization," "the building of a well-off society," "poverty alleviation," "land transfer" and "Strategy for Vitalizing Villages" put forward by the Chinese government will surely trigger greater changes in economic layout and development needs and pose higher requirements for water resources utilization and protection, which will definitely need water resources allocation to be in line with development, and need to protect water resources while developing economy, and also need a more rational, precise and intelligent water resources management model.

Second, water resources will face special demands that come along with the "Belt and Road" Initiative, the building of "Yangtze River Economic Belt," the "Coordinated Development for the Beijing-Tianjin-Hebei Region," the construction of "Xiong'an New Area" and of new economic development zones in various cities. In order to promote economic and social development, form a new pattern of opening-up in an all-round way, implement the strategy of regionally coordinated development and forge a modern economic system, the Chinese government has put forward a series of strategies, initiatives and development plans, which pose new demands for water resources utilization and protection. The original spatial distribution of water resources and the pattern of water resources allocation cannot fully stand up to the changes of these major strategies and constructions; therefore, they need to be re-studied and demonstrated so as to achieve the scientific allocation and rational utilization with water saving as the priority and protection as the major content.

Third, the rationalization and self-reform of water resources administrative organs and management systems as a result of reform and opening-up have promoted the development of water resources–related disciplines and expertise. Reform and opening-up target not only the economic field but also management organizations, institutions and system. In order to adapt to reform and opening-up and the needs of economic and social development, water resource management institutions and system should, on the one hand, be continuously streamlined. For example, in order to better respond to the institutional reform proposed by the State in 2018, water resources management institutions should be streamlined in a timely manner to better serve economic and social development as well as people living and working in peace and contentment. For example, in order to follow closely the strictest water resources management system and the river chief system proposed by the State, the unified management system of water resources should be adjusted in time so as to guarantee the water resources utilization and protection. On the other hand, self-reform is necessary to continuously improve water resources management institutions and systems. For example, China carried out the river chief system, the reform of the unified management institution of natural resources, the strictest water resource management system, water rights trading system, agricultural water price reform and subsidy system, so as to continuously enrich the work and the content of discipline related to water resource utilization and protection.

3.2. Innovation

Innovation is the unique cognitive and practical ability of human beings, the driving force for national progress and social development, and the source of vitality for the formation of innovative thinking, research and development of new products, scientific and technological progress, and the improvement of management efficiency and service quality. The report to the Nineteenth CPC National Congress pointed out: "Innovation is the primary driving force behind development; it is the strategic underpinning for building a modernized economy." "We must pursue with firmness of purpose the vision of innovative, coordinated, green, and open development that is for everyone." "We must ensure our theory evolves with the times, deepen our appreciation of objective laws, and advance our theoretical, practical, institutional, cultural, and other explorations." To build an innovative country, it put forward that China will have become a global leader in innovation by 2035. The construction of an innovative country raises new demands for water resources utilization and protection, and at the same time brings vitality and development opportunities.

(1) Innovative development lays down higher requirements for water resources utilization and protection. Along with population growth and economic expansion, the contradiction between water resources supply and demand is prominent. Water resources security and ecological protection face more severe challenges. We can only realize the balanced development of water resources utilization and protection under the background of innovative development by continuously innovating water resources management modes, changing traditional concepts governing water resources management, innovating technology in water resources utilization, improving water efficiency and striking the balance between water resources supply and demand.

(2) The construction of an innovative country itself includes water resources utilization and protection-related innovation in theories, technologies, practices, institutions and cultures. Water resource is indispensably basic for human survival and development. Therefore, its utilization, protection, innovation and construction must be leading our times in order to support the creation of an innovative country. For example, technologies for flood prevention and drainage, water saving, sewage treatment and water resource management, along with water resource management mode and system and the construction of system platform and intelligent

water supply are all important parts and supportive foundation of the building of an innovative country.

3.3. Green development

China's economy made great achievements in the first thirty years of reform and opening-up, which also brought serious consequences such as the over-consumption of resources, environmental pollution and ecological deterioration, all threatening our living environment, beautiful homeland and physical health. In order to reverse this situation, it is necessary to transform the current economic development mode into one with green development as the core.

Green development is a new-type sustainable development model based on the traditional development concept. It emphasizes the constraints of ecological and environmental capacity and resource-carrying capacity and takes environmental protection as the priority. Only by vigorously promoting green economic development can we both develop the economy and protect the environment. Green development has become the mainstream trend of development of the world today. The report to the Nineteenth CPC National Congress pointed out that "the entire Party and the whole country have become more purposeful and active in pursuing green development," to "develop eco-friendly growth models and ways of life," "cultivate ecosystems based on respect for nature and green development" for "promoting green development." Green development itself requires sustainable, rational and protective utilization of water resources. Therefore, it raises higher requirements for water resources utilization and protection.

(1) The sustainable utilization of water resources is an essential condition of green development. Water resources are the supporting condition of economic and social development and the important material basis of production, life and ecology. Green development is possible only with the sustainable utilization of water resources, which is the prerequisite for implementing the concept of green development.
(2) Green development requires good use of water resources. The fresh water resources that can be renewed in a certain region and a specific period of time are limited. In order to pursue economic development without destroying water resources, we must make good use of water resources. For one thing, we need to curb the unrestrained demand, implement the water-saving priority strategy and control the total amount of water consumption. For another, we must improve the allocation of water resources and the use of water resources, and give full play to the comprehensive benefits of water resources. In addition, we need to increase pollution control efforts and reduce pollutant emissions. Reducing the originally

recyclable pollution to the water resources system means, in fact, indirectly increasing the amount of water resources available. At the same time, we should fully develop and utilize unconventional water resources, such as reclaimed water, rainwater, desalinated sea water and more.

(3) The efficient utilization system of water resources should be included in the green development hierarchy, which comprises green production and consumption, green, low-carbon, and circular development, green finance, green innovative technology, energy saving and environmental protection industry, clean production industry, clean energy industry, resource saving and recycling, green and low-carbon ways of life, green family, green school, green community and green travel, and so on. These include efficient utilization of water resources, saving of water resources, water-saving and environmental protection industries, water-saving actions, the building of water-saving society, water-saving enterprises, universities, communities and families.

(4) The sustainable utilization of water resources would not be possible without green development being put in place. However, to achieve the sustainability of water resources is a systematic project, which calls for actions from multiple aspects such as water use concept, project investment, legal system, policy orientation, economic means, technical system, industrial development, life style and consumption concept.

3.4. The advancement of ecological conservation

Ecological civilization is a form of social civilization that rises after primitive civilization, agricultural civilization and industrial civilization. Ecological civilization is a "civilized stage" in the history of human development, a very long one indeed. The building of ecological conservation is just the process of turning this stage of civilization into reality. On November 8, 2012, the report to the Eighteenth CPC National Congress put forward that the construction of ecological conservation is a long-term plan related to the well-being of the people and the future of the nation, and should be put in a prominent position. This is the inevitable choice when we face the severe situation of tightening resource constraints, serious environmental pollution and ecosystem degradation. The safety of water resources is such an important guarantee for the advancement of ecological conservation, which in turn lays down higher requirements for the utilization and protection of water resources.

First, water resource is the core constraining factor in the building of ecological conservation, as well as the most essential basis and critical element of ecological civilization. Water is the source of all life, the controlling element of ecology and environment, the essential substance in human life and production, and the soul of civilization. It is an important foundation for promoting ecological conservation and the harmony between human and nature to carry on water resources' rational development, optimized configuration, saving-oriented utilization, effective protection and scientific management through water engineering for the purpose of safe water use and enhancing the ability of water resources to guarantee the sustainable development of economy and society.[11]

Second, water resource conservation and water ecological protection are central aspects of ecological conservation. Saving resources is a fundamental way to protect environment, and a good environment is essential to the sustainable development of human society and economy. The direct goal of ecological conservation is to protect the ecology and environment on which mankind depends. Saving water resources is a major way to address water shortage, while protecting water ecology constitutes an integral part of ecological conservation. Both are principal measures to advance ecological conservation featuring the harmony between human and nature.

Third, sustainable utilization of water resources is imperative for ecological conservation, for which, we must toughen the management of water resources, ensure the sustainable use of water resources and strengthen the basic functions of water resources in ecological civilization. Otherwise, the ecological environment will not be promoted, let alone ecological conservation.

3.5. Protecting ecological environment

Ecological environment refers to the physical environment in the organic relationships between living organisms, on which human existence depends. It includes ecological factors that are either biological or non-biological. Natural and geographical conditions like plants and vegetation, rivers and mountains, land and climate, along with man-made environment, constitute the basic conditions for human survival and evolution.[12] The report to the Nineteenth CPC National Congress reiterated environmental protection by saying that "we have a long way to go in protecting the environment," and we must "cherish the environment as we cherish our own lives. We will adopt a holistic approach to conserving our mountains, rivers, forests, farmlands, lakes and grasslands, implement the strictest possible systems for environmental protection," and "we must create good working and living environments for our people and play our part in ensuring global ecological security."

Water is a vital component of environment and one key to the sound operation of the ecosystem. The quality of environment is closely related to the quantity and quality of water available. Due to the fact that water resources in nature are limited, an increase in water consumption by one man means the decrease on the part of another one. It is especially true that the ignorance of the nature's demand for water might easily cause the river to shut down, the lake to dry up, the wetland to shrink, the soil to be salinized, grassland to degrade, forest to be destructed and sand to desertification. This will seriously restrict economic and social development and even destroy human living environment. Therefore, environmental protection has laid down clear requirements for water resources utilization and protection.

First, a major part of environment protection is the protection of water resources. Water resource is an integral part of environment and an irreplaceable element in ecosystem and natural environment. The type of water resources in a place decides and classifies its environment. Therefore, water resource protection is the main content of environmental protection.

Second, the ecological environment protection proposed in the new era puts forward higher requirements for water resources utilization and protection. In the new era, the construction of socialism with Chinese characteristics puts forward higher expectations for ecological environment protection, so it also puts forward higher requirements for water resources utilization and protection, including ecological water use, ecological basic flow, ecological water level and river runoff process, water resources allocation, water environment, etc.

3.6. Realizing the Chinese dream

The Chinese Dream is an epochal guiding ideology and ruling idea put forward by the Eighteenth CPC National Congress, as the great rejuvenation of Chinese nation is our most cherished dream since modern times. The report to the Nineteenth CPC National Congress pointed out that "the Chinese Dream of national rejuvenation will be realized ultimately through the endeavors of young people, generation by generation."

The Chinese Dream is mainly about the future of China and embodies our people's hopes and expectations for the great rejuvenation of the Chinese nation. It is the dream that the whole Chinese nation will pursue unceasingly, a long-cherished wish that hundreds of millions of people will pass on from generation to generation. Every Chinese is the participant and the creator as well of the Chinese Dream.

The blueprint depicted in the Chinese Dream envisions a country where people are vital, down-to-earth, pioneering, innovative and dedicated; where the nation is prosperous and rejuvenated with happy people, harmonious society and a beautiful homeland. But, the Chinese Dream in the first place means to protect the lucid water and lush mountains of the motherland, and that people have established and are practicing the concept that the lucid water and lush mountains are invaluable assets.

First, the protection of water resources is the precondition and important content of the Chinese Dream. As an irreplaceable resource for human existence and development, water is the source of life, the necessity of production and the basis of ecology. To protect the water resources means to protect the environment and human beings. Environmental protection is an integral part of and a crucial material basis for the Chinese Dream.

Second, the Chinese Dream poses higher requirements for water resources protection, for it is a grand blueprint that includes all aspects involved in natural resources protection and the advancement of human society. Economic development, social progress and ecological protection are all inseparable from the support of water resources, which lays down stricter requirements for water resources' scientific and reasonable utilization and optimal allocation as well as for the protection of water environment.

4. PROSPECT OF WATER RESOURCES UTILIZATION AND PROTECTION AS DEMANDED BY THE DEVELOPMENT OF CHINA IN THE NEW ERA

According to the foregoing analysis of the development course of water resources utilization and protection in the past forty years of reform and opening-up in China, and by considering the needs of the construction of socialism with Chinese characteristics in the new era, water resources utilization and protection, as the author believes, are still at the protection-focused stage, which is expected to continue up to 2025. The next stage, according to our judgment, will be the stage of intelligent water use, transforming to which needs full access to modern information and communication technology and cyberspace virtual technology. In order to better serve the construction of socialism with Chinese characteristics in the new era, Chinese government has come up with a series of strategic measures, such as the Belt and Road Initiative, the Yangtze River Economic Belt Strategy, the coordinated development for the Beijing-Tianjin-Hebei region, the construction of Xiong'an New Area and the Strategy for Vitalizing Villages, etc. These strategic deployments will be new areas in which water resources utilization and protection can play a role. Based on the above analysis, we present a prospect on the water resources utilization and protection that can satisfy the development needs of the new era in China, as shown in Figure 5-1.

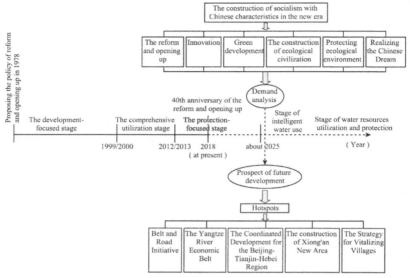

Figure 5-1 Water Resources Utilization and Protection in China from 1978 to 2018 and Prospect of Future Development

4.1. Prospect of the new stage of water resources utilization and protection

4.1.1. Protection-focused stage is expected to continue around 2025

According to the above analysis, in the forty years of national construction since the reform and opening-up in 1978, China has achieved remarkable results in water resources utilization and protection and has gone through three stages that, respectively, focused on development, comprehensive utilization and protection. Among them, the protection-focused stage began in 2013, when the country vigorously advocated the building of ecological conservation which includes the development of water ecological conservation, with special emphasis laid on the status of water ecological conservation and the goals of building ecological civilization. This is the demand for water resources utilization and protection posed in China's current socialist drive, as well as the new era's guiding thinking, which, according to the author's analysis, will continue to play an important role. Currently, the protection-focused stage will last at least until 2025. Then, the trend of deterioration of water ecology will have been basically reversed and the water resources safety effectively safeguarded. The policy system for water resources utilization and protection will have been basically straightened out, which is beneficial to ecological conservation. China will have basically solved some difficult problems related to water, including water resources' utilization planning, protection technology, system automatic monitoring and optimization management, and so on.

Based on the analysis of the development goals and research progress of this stage, the author believes that water ecological conservation and the building of ecological civilization will be the major goals of this protection-focused stage, with special emphasis laid on questions related to water resources utilization and protection as follows.

First, basic theory researches support the building of ecological civilization and water ecological conservation, including the influence and action mechanism of ecological engineering construction, ecological hydrological process and mechanism, biological diversity action mechanism and analysis methods, hydrological threshold and mechanism that maintain ecosystem stability; the mechanism and simulation of the pollutant migration process and accumulation process, water ecological redline theory and method, calculation method for ecological water demand, ecological hydrology, urban hydrology, ecological engineering research, etc.

Second, research, development and practice of technologies for water pollution control and ecological restoration. Water pollution control technologies include artificial aeration with air or pure oxygen, falling water aeration, biofilm, ecological landscape pond, artificial wetland, biological floating bed, ecological ditches and so on. Water ecological restoration technologies are mainly about the construction of river sinuosity, the restoration of river cross-sectional diversity, in-river habitat construction, fish-passing facilities construction, water system connection, river-lake shoreline control, ecological slope, water landscape construction, ecological dredging, etc.

Third, comprehensive water-saving technology and the construction of a water-saving society, which include technologies for agricultural, industrial and domestic water-saving, sewage and wastewater regeneration and reuse, rainwater utilization, seawater desalination, automatic monitoring of comprehensive water-saving, and precise management system, the planning, design and action of water-saving society, etc.

Fourth, theories and methods for the harmonious development of water resources, economy and society, which include theories, methods, regulation models, the generation and evaluation of regulation schemes, optimization and allocation of water resources based on harmonious development, comprehensive water resources planning methods, river-lake system connection and theories for water engineering optimization and layout.

Fifth, water resources management-related policy system and institutions adapted to the building of ecological civilization, which include the technical outline and guideline of ecological conservation (in the field of water resources), the legal system and institutional system of water resources management that meet the development needs of the new era, the water resources management system tailored to the reform of national institutions, and the water ecological compensation system, water rights, water price and water market construction, etc.

4.1.2. The stage of intelligent water use will be ushered in (expected to be after 2025)

To predict what the next stage will be with regard to water resources, we need to take into consideration the development trend of water resources utilization and protection in China, and the country's actual needs in its national construction in the new era as well as the driving force from the "Internet +" development, especially the influence from information and communication technology and the virtual cyberspace technology. According to our judgment, the next stage should be the stage of intelligent water use, which is expected to come after 2025.

Based on abundant experience in water resources utilization and protection, this new stage will make full use of information and communication technology as well as the virtual cyberspace technology in the transformation from the traditional mode of water resources utilization and protection to the intelligent one. On the one hand, the modern information and communication technology and cyberspace technology which developed since the end of the twentieth century has laid a solid foundation for the transformation. On the other hand, since the mid-twentieth century, especially since China's reform and opening-up in 1978, rich results have been produced in the theories, methods and practice of water resources utilization and protection. On top of the experience gained in recent years and until 2025, or the protection-focused stage, China should have been empowered with profound achievements and rich experience in hydrology, water resources, water environment, water safety, water engineering, water economy, water law and water culture. The knowledge reserve in this stage can be regarded as a preliminary preparation for the stage of intelligent water use. China is expected to enter a new era built on "intelligent water use" by 2025. Referring to the analysis of "intelligent water conservancy,"[13] we describe the outline of the stage of intelligent water use as follows:

First, water resources utilization and protection call for making full use of information and communication technology and virtual cyberspace technology and demonstrates high intelligence.

Second, China will have realized, for the water resources, the automatic monitoring, digitalized data, quantified models, intelligent decision, information-based management and standardized policy system.

Third, China will have established a basic platform for the internet of water, which integrates the physical water network of rivers and lakes, the virtual water network of spatial and three-dimensional information, and the distribution network of water supply, water use and drainage. Then, China will have functionally integrated system that can realize real-time monitoring, rapid information transmission, accurate water condition forecast, decision-making for service optimization, accurate water quantity allocation and the comprehensive management of water resources.

According to the general description of this stage, the author thinks that the focus of research on water resources utilization and protection in the stage of intelligent water use will be on the following areas.

First, information and communication technology as well as virtual cyberspace technology will be further promoted, applied and studied in the field of water resources utilization and protection. This is the key and landmark technology of the intelligent water use, which will be used to construct the basic platform for the internet of water that integrates physical, virtual and distribution water networks.

Second, the building of the intelligent brain for real-time monitoring of water resources, large data transmission and storage technology, fast computing based on cloud technology, intelligent water decision-making and intelligent water scheduling research.

Third, the execution service system for intelligent water use. Based on communication technology and virtual technology, this execution system is constructed to realize "real-time monitoring of water resources, rapid information transmission, accurate water condition forecast, decision-making for service optimization, accurate water quantity allocation and the comprehensive management of water resources." As it integrates all these functions, it can provide customers with tailored order service and accurate delivery at any time, which involves water cycle simulation, efficient utilization of water resources, flood prevention, drought relief and disaster reduction, water environment protection, water safety guarantee, the scientific planning of water engineering, water rights trading, the construction of water legal policy system, and the inheritance and construction of water culture, among many other needs.

4.2. Prospective research hotspots for water resources utilization and protection

4.2.1. The Belt and Road initiative

The Belt and Road Initiative, or the "Silk Road Economic Belt" and the "21st Century Maritime Silk Road" or "B&R" in short, is an open and world-oriented cooperation initiative proposed by the Chinese government. The Belt and Road, which spans Asia, Europe and Africa, aims to promote exchanges and cooperation among countries along the routes by giving full play to the roles of policies, facilities, trade, funds and people's will. The initiative advocates the awareness of a "community with a shared future for mankind" and the common development of all countries. Thus, it is the top-level design and the responsibility of China as a big power.

Most of the countries along Belt and Road Initiative are lower-middle-income countries. Compared with those of relatively high per capita income in Central and Western Europe, those in North Africa, East Africa and South Asia are of lower per capita income. It is especially true with some African countries. In addition, most "B&R" host countries have bottlenecks in their economic and social development, with water problems, such as flood, drought, water shortage, water pollution, insufficient water supply and so on, being the most prominent. The most prominent water problems in different regions are different, so are the difficulties in addressing them.[14] Water security underlies the smooth implementation of the "B&R" initiative and the common and sustainable development of all countries; therefore, the research on water resources utilization and protection in countries along the "B&R" routes will surely be the focus of attention in the future.

Based on the analysis of the water resources problems in "B&R" host countries and on the judgment of the future development demand, the paper presents the future research focuses and key work of water resources utilization and protection as follows.

First, the risk assessment system for region-based water resources along the "Belt and Road." The "B&R" countries traverse a wide range of areas known for huge differences in water resources, economic status, social backgrounds, especially ethnic customs. What comes along is the drastic diversity in the categories of water resources disasters and safety emergencies and in the abilities to battle the risks brought by these disasters and incidents. Therefore, the establishment of a sound risk identification and assessment system requires not only the assessment of the safety status of water resources in countries along the Belt and Road route but also a dynamic appraisal of the potential risks and their ability to fight such risks.

Second, the establishment of a regional water resources allocation network system. The "Belt and Road" participating regions display sharp disparities in water resources utilization and supply, highlighting the imbalance between water demand and stock, which cannot be corrected without a rational allocation. The construction and implementation of the projects of interconnecting river and lake water systems, for example, is an engineering answer to the problem. A regional water resources allocation system can be established gradually if we relocate water resources by resorting to the river and lake interconnectivity program.

Third, the research on the spatial balance in the coordinated development of water resources, society and economy. The implementation of "B&R" Initiative has expanded the communication in the areas of culture and technology and gradually narrowed the regional and national disparities, which renders an equilibrium in the development of water resources, society and economy. Therefore, it is in accord with the initiative of "building a community with shared future for mankind" to strike a spatial balance of the development in different regions, thus making due contribution to the people in the countries along the route.

Fourth, the establishment of security monitoring and early warning mechanisms for water resources in countries and regions along the "Belt and Road." The many insecure factors endangering the security of the water resources along the route, as well as the frequent occurrence of water resource disasters, have made it imperative, for the smooth implementation of "B&R" Initiative, to strengthen security surveillance and early warning so as to monitor in real time the water resource calamities that might happen. In doing so, we can prevent or reduce, to a large extent, the occurrence of water resources disasters.

Fifth, the exchange, promotion, R&D and application of the technologies for the development, utilization, saving, protection and treatment of water resources in "B&R" participating countries. It is multilaterally beneficial for all countries to extend their communication and cooperative R&D for common prosperity, for many of them are at different stages of national and technical development and some technologies are highly complementary. This is one of the major intentions of China's advocation of "B&R" Initiative.

Sixth, the construction of a mechanism for the scientific exchange, cooperation and sharing for water resources. Through bilateral cooperation, departments, universities, research institutions and laboratories can train talents by exchanging students, and carry out international collaboration to enhance cultural, scientific and technological interactions, so as to raise public awareness of cherishing and protecting water, improve the efficiency of water resources utilization, protect water resources and provide a security guarantee for "B&R" Initiative.

4.2.2. The Yangtze River Economic Belt

The Yangtze River Economic Belt will remain the focus of attention for a foreseeable future. With superior natural conditions and a higher development level than adjacent regions, the Yangtze River basin has played an important role in China in history and today. The economic circle along the Yangtze River is called the Yangtze River Economic Belt, which covers eleven provinces and municipalities including Shanghai, Jiangsu, Zhejiang, Anhui, Jiangxi, Hubei, Hunan, Chongqing, Sichuan, Yunnan and Guizhou, with an area of about 2.05 million square kilometers. As one of the regions with the strongest comprehensive strength and that provide largest strategic support in the country at present, it is the economic corridor of interaction and cooperation between the East, Central and West China.

The Yangtze River Economic Belt is designed on the basis of the Yangtze River which, thanks to its rich water resources, boasts large carrying capacity. However, with the rapid development of the economy and society, the Yangtze River has seen sharply rising water consumption, pollution and eco-degradation, which has resulted in severe restriction to the regional societal and economic progress in the new period. In order to address this dilemma, the Chinese government has put forward the policy of "well-coordinated environmental conservation and avoiding excessive development," which can be translated into low-carbon and green development. Currently, in the face of grave challenges, it is of great significance to seek a harmonious development of the Yangtze River Economic Belt, which, means to coordinate protection and development in the first place and balance the water resources utilization and protection in the second place. At the same time, it is also faced with severe challenges, which will be the focus of attention for a period of time in the future.

Based on the analysis of the current water resources problems and demand in the Yangtze River Economic Belt, the chapter presents future research hotspots and relevant key work as follows:

First, basic theory researches on water resources utilization and protection, which meet the Yangtze River Economic Belt's development needs or address academic frontier questions. The Yangtze River Basin has always been one of the key areas in China for exploitation, the degree of which and the level of economic development are relatively high. However, it also brings a series of problems, the core of which is the obvious tendency of the overexploitation of resources and the deterioration of environment. In order to reveal the relationship between and the feasibility of development and protection, it is necessary to further study the interaction, mechanism and evolution law that exist in the relationship between human social activities and natural resources system to provide a basic research basis for water resources utilization and protection.

Second, the theory of harmonious protection and development and its application in practice. The dilemma faced by the Yangtze River Economic Belt is stemmed from the inconsistency between protection and development, which are often regarded as the opposite pairs of contradictions. Development causes problems and influences protection and vice versa, which might put economic benefits in jeopardy. Therefore, the simultaneous needs for development and protection involve the question of coordination, which means the theory of harmonious development might find a good application here. The theory for the harmonious development and protection of the Yangtze River Economic Belt can be developed gradually upon the existing theories and practices before being put into test in practice.

Third, the theory and practice of ecological conservation in the Yangtze River Economic Belt. The vigorous promotion of ecological conservation constitutes the basis and guarantee for the "well-coordinated environmental conservation and avoiding excessive development" in the Yangtze River Economic Belt, as well as the focus of work in the green development and ecological conservation in key areas in China. Therefore, it is necessary to explore how to develop the theory of ecological conservation specific or unique to the Yangtze River Economic Belt and have it tested in local practice.

Fourth, the technology innovation and application that can promote the efficient utilization of water resources and sustainable economic and social growth. The development of technologies for water saving, use, ecological conservation and restoration, scientific distribution and intelligent water network, among many others, can enhance China's competence in water conservation and the efficiency of water resources utilization, which will facilitate an economic development pattern based on the protection of water resources.

Fifth, the research on and implementation of the strictest water resources management system and the river chief system. In order to implement the policy of development and the protection of the Yangtze River Economic Belt, it is a must to formulate a series of strict management systems, including the strictest water resources management system and the river chief system, which is the premise and system guarantee for harmonious development and ecological conservation.

4.2.3. The coordinated development for the Beijing-Tianjin-Hebei region

As a major strategy of China, the coordinated development for Beijing-Tianjin-Hebei means a holistic and balanced growth of the three places. The core of the strategy is to build a new modern capital circle by shifting out the functions that do not belong to Beijing as the capital city so to treat the "big city disease" and adjust and improve the urban layout and spatial structure. On April 30, 2015, the Political Bureau of the CPC Central Committee reviewed and approved the *Outline of the Plan for Coordinated Development for the Beijing-Tianjin-Hebei Region*, a top-level design for the comprehensive development of the region, which has clarified and detailed the rules and road map of the coordinated development for the Beijing-Tianjin-Hebei region.

The coordinated development for the Beijing-Tianjin-Hebei region has an important demonstration effect on the sustainable development of city clusters in China, for the building of a future-oriented capital city economic circle can give full play to the complementary advantages of the Beijing-Tianjin-Hebei region while improving and optimizing the layout of urban clusters. Investment might flood in, changing the industrial pattern of the Beijing-Tianjin-Hebei, as well as the coordinated development of the northern hinterland of the region. Being the political, economic and cultural center of China, the region is known for its high population density, rapid economic development, short supply of resources, inadequate environmental capacity and outstanding contradiction between development and protection. How to coordinate the water resources utilization and protection has always been a key issue of the region, and will remain the focus of attention in the future.

Based on the analysis of water resources problems and the future development demand that might arise in the coordinated development for the Beijing-Tianjin-Hebei region, the chapter prospects the research hotspots and key work related to the utilization and protection of water resources in the three places.

First, the optimal allocation of water resources and the construction of river-lake connection system for the coordinated development of Beijing-Tianjin-Hebei region. Judging from the real needs from the balanced development of Beijing, Tianjin and Hebei, China should give consideration to water resources like local water, water from the South-to-North Water Diversion Project, desalinated seawater and reclaimed water, and study the optimal allocation of water resources in accordance with its spatial matching with land resources. The country should build a river and lake connection system that can satisfy the needs of intelligent water conservancy, which can improve China's water resources regulation competence and water supply capacity so that the river and lake systems can serve the purpose of "water storage and discharge, regulation of high flow and low flow, water diversion and drainage, water source complementation, and ecological health."

Second, it is necessary to conduct the integrated planning, allocation, dispatching, protection, monitoring and assessment of water resources in Beijing, Tianjin and Hebei to support the coordinated development of the region.

Third, an integrated water-saving society in Beijing, Tianjin and Hebei should be built by implementing the policy of prioritizing water conservation. The region is short of water, with sharp contradiction between water demand and supply. The first choice, therefore, is to save more water, from which we can find a way out and improve efficiency so as to build the region into an integrated water-saving society as soon as possible.

Fourth, an integrated water network for the Beijing-Tianjin-Hebei region. Beijing-Tianjin-Hebei region should be built. As the political, economic and cultural center of China, Beijing-Tianjin -Hebei region boasts a high level of economic development and enough capacity and conditions to build an integrated intelligent water network. Therefore, it should take the lead in implementing intelligent water conservancy strategy from the national level as a demonstration area.

Fifth, carrying out studies on water resources management system should be carried out under the objectives of ecological conservation and green development. On the one hand, it is imperative to adapt to the development needs of the new era by strengthening ecological protection with green development. On the other hand, it is urgent to establish a water resources management system which can meet these development needs, and create and improve an institutional system that can support and guarantee the success of the development road designed for the new era.

4.2.4. The construction of Xiong'an New Area

On April 1, 2017, the CPC Central Committee and the State Council announced the decision to set up Xiong'an New Area, a state-level new area after Shenzhen Special Economic Zone and Shanghai Pudong New Area, which could be looked upon as a major historic strategy made by the CPC Central Committee. Located in Baoding City, Hebei Province, Xiong'an New Area lies in the hinterland of Beijing, Tianjin and Baoding, covering three small counties, namely, Xiong, Rongcheng and Anxin and surrounding areas.

In response to the coordinated development for the Beijing-Tianjin-Hebei region, the construction of Xiong'an New Area is of great significance to the orderly migration of Beijing's non-core functions, the adjustment and optimization of urban layout and spatial structure, the cultivation of innovative driving forces and the development of new engines. As a state-level new area to be built in the new era, which will naturally set a high starting point and standards, it will fully follow the new ideas of CPCCC and central government in governing the country, such as the theories for building a green, ecological and livable area, or one driven by innovation, reform and opening-up, or one with a coordinated social, economic, environmental and resources development. Located in North China Plain with poor water resources endowment and prominent contradiction between water supply and demand, Xiong'an New Area, if to be built into a super-large urban area with guaranteed sustainable water supply, needs badly a scientific and rational utilization of local water, externally transferred water and non-conventional water, thus will surely attract the attention of researchers for a period of time in the near future.

Based on the above analysis of water resources problems and future development demand in Xiong'an New Area, the focus and related key work of water resources utilization and protection in the area are prospected as follows:

First, the basic theoretical researches on water resources that meet the needs of new era, such as those for reform and opening-up, innovation-driven and green development, ecological civilization, the realization of the Chinese Dream, among others. Xiong'an New Area used to have a low level of development but is now to be built into an urban area with rapid high-intensity activities. The original water resources system and ecosystem will undergo changes in the extent and speed rarely seen in other areas. Therefore, in Xiong'an, an area to witness fast expansion as well as the new needs of our times, it is necessary to strengthen the basic researches on water resources theory, including the evolution mechanism of water resources system and ecosystem, the influence mechanism of human activities, the scientific regulation mechanism, the threshold value of water resources utilization and of ecosystem.

Second, the construction of an intelligent water network with optimized allocation of water resources that feature multiple water resources, multi-objectives and strict demands, which can meet the needs of new era. For this purpose, it is necessary to study the methods and schemes for optimal allocation of water resources featuring multi-water sources, multi-objectives and strict-demand in Xiong'an New Area, which is under the rapid and large-scale disturbance of human activities. Then, we can study and formulate the plan for the construction of intelligent water network based on the findings of the above researches.

Third, the R&D and application of high technologies for water resources utilization and protection. As a demonstration new area led by innovation, Xiong'an New Area needs the R&D and application of high techs for water resources conservation, utilization and protection to improve efficiency and promote its efficient, rapid and green development.

Fourth, the construction of a water resources management system that meets the special development needs of the state-level new areas. The water resources management system in general regions or basins does not necessarily meet the high standards and requirements of Xiong'an New Area, which means that it is necessary to formulate a new water resources management system according to the positioning, functions and constrictive conditions of the new area. The new system should include water index control, requirements for water conservation, water pricing system, initial water rights distribution, water rights trading, sewage rights distribution and trading, water resources protection laws and so on.

4.2.5. Strategy for vitalizing villages

On October 18, 2017, the report to the Nineteenth CPC National Congress put forward the "implementation of the strategy for vitalizing villages." The No. 1 document released in 2018, namely, the *Opinions of the Central Committee of the CPC and the State Council on Implementing the Strategy for Vitalizing Villages*, laid down the schemes for agricultural and rural development in an all-round way, setting timetables: by 2020, significant progress will be made in rural revitalization, and the institutional framework and policy system will be basically formed; by 2035, agricultural and rural modernization will be basically realized; by 2050, rural areas will be revitalized in an all-round way, with strong agriculture sector, beautiful countryside and full realization of farmers' wealth, to ensure local farmers access to long-term stable income and happy livelihoods.

China is a large agricultural country. In 2017, the permanent population of rural areas accounted for 41.48% of the total population of the country. As the primary industry, agriculture has to produce food for more than 1.3 billion people in the country. The problem of "agriculture, rural areas and farmers" is a fundamental one related to the national economy and people's livelihood, as well as a key point for poverty alleviation and building a moderately well-off society in an all-round way. Agriculture is also the largest water consumer in China. In 2016, the water consumption by agriculture in China accounted for 62.4% of the total water consumption. Therefore, the Strategy for Vitalizing Villages will inevitably involve a series of water-related reforms, innovation and development issues, which will remain the focus of public attention for a period of time.

Based on the above analysis of the water resources and future development demand involved in the Strategy for Vitalizing Villages, the research hotspot and related key work of water resources utilization and protection in the Strategy for Vitalizing Villages are prospected as follows.

First, the integrated construction of water resources against the background of comprehensive urban and rural development. Rebuilding the urban-rural relationship and realizing the integrated urban-rural development are the major trends in implementing the Strategy for Vitalizing Villages, and the integrated construction of water resources in urban and rural areas is the need of unified management of water resources.

Second, policy choices and development approaches for coordinated development of water resources and land resources. The reform of water resources utilization policy is closely related to land reform, with water use efficiency, water pricing system and water rights trading all related to land use policy. The choice of policies and approaches suitable for rural development in respective regions is an important factor for the realization of Strategy for Vitalizing Villages.

Third, the utilization and protection of water resources needed for green development and ecological conservation in rural areas. The rural development in the Strategy for Vitalizing Villages must be green one and pass on agricultural civilization on the road of ecological conservation. We should choose, in this context, the working ideas, schemes and safeguard measures that are in the best interest of water resource utilization and protection.

Fourth, the construction of water resources management system that meets the needs of rural governance system, precise anti-poverty efforts and poverty reduction with Chinese characteristics. The new era demands the building of a new rural water resources management system, which includes water conservation system, the strictest water resources management system, river chief system, farmer water utilization association system, water pricing system, water rights trading system and so on.

5. CONCLUSION

This chapter divides water resources utilization and protection in the past forty years of reform and opening-up in China into three stages: the development-focused stage (1978–1999), the comprehensive utilization stage (2000–2012) and the protection-focused stage (2013–2018), with the last stage to continue to the year of 2025. The chapter, under the guidance of the thoughts on socialism with Chinese characteristics in the new era and from the perspectives of reform and opening-up, innovation, green development, ecological conservation, protection of ecological environment and realization of the Chinese Dream, systematically analyses the country's new demands for water utilization and protection in the new age. The chapter comes up with the judgment that China will enter the period of "intelligent water use" (which is expected to be after 2025), and presents a description of the development framework for this period of time. According to the analysis, information and communication technology, as well as virtual cyberspace technologies will be fully employed in intelligent water use, which takes smart water network as the carrier and the intelligent water use as the main form. The chapter presents, on the basis of the analysis of demand and development stage, five future hot areas of water resources utilization and protection, namely, the Belt and Road Initiative, the Yangtze River Economic Belt, the Coordinated Development for the Beijing-Tianjin-Hebei Region, the construction of Xiong'an New Area and the Strategy for Vitalizing Villages, putting forward the research focuses and key work of water resources utilization and protection as a reference for the future.

This chapter only makes analysis and judgment on the water resources utilization and protection during the forty years of reform and opening-up as well as the development trend in the future. Being macro and preliminary, it might not be as comprehensive and concrete as expected, and might even be controversial in some judgments that need further researches from more scholars, who can also provide support for reform and opening-up, innovation, green development, ecological conservation and the realization of the Chinese Dream

Notes

1 The data are from China Statistic Yearbook 2016.
2 The data are from China Water Resources Bulletin 2016.

3 Xia Jun, Zuo Qiting and Shen Dajun (2018). *The Science of Water Resources*, China Science and Technology Press.
4 Cao Xingrong (2010). Brief Discussion on the New Stage of Water Resources Development and Utilization, *Water Resources Development Research*, No. 4.
5 Qu Yaoguang, Ma Shimin and Liu Jingshi (1995). Development and Utilization of Qater Resources in Northwest China. *Journal of Natural Resources*, No. 1.
6 Zhu Meiling (2002). The Development and Utilization of Water Resources in the Habahe River Basin in Xinjiang and the Appropriate Water-Saving Irrigation Ways. *Territory and Natural Resources Study*, No. 3.
7 Wang Yahua, Huang Yixuan and Tang Xiao (2013). The Division of Water Resources Development Stages in China: Theoretical Framework and Evaluation, *Journal of Natural Resources*, No. 6.
8 Zuo Qiting (2015). China's Water Resources Development Stage and future "Water Resources 4.0" Strategic Vision, *Water Resources and Power*, No. 4.
9 Chen Lei (2008). Carrying on the Past and Opening up and Advancing with The Times at the New Historical Starting Point to Promote the Sound and Rapid Development of Water Conservancy – Speech Addresses to the Cadres at the Conference of the Ministry of Water Resources to Commemorate the 30th Anniversary of Reform and Opening up, *China Water Resources*, No. 24.
10 Zuo Qiting (2009). Harmony between Human and Water Resources – From Concept to Theoretical System, *Water Resources and Hydropower Engineering*, No. 8.
11 Zuo Qiting (2013). Discussion on Several Key Issues of Water Ecological Civilization Construction, *China Water Resources*, No. 4.
12 Zuo Qiting and Wang Zhonggen (2006). *Modern Hydrology* (2nd edition), The Yellow River Water Conservancy Press.
13 Zuo Qiting (2015). China's Water Resources Development Stage and future "Water Resources 4.0" Strategic Vision, *Water Resources and Power*, No. 4.
14 Zuo Qiting, Hao Lingang, Ma Junxia and Han Chunhui (2018). Thoughts on Water Problems in "Belt and Road" Zones and Reflections on Using Chinese Water Control Experience for Reference, *Journal of Irrigation and Drainage*, No. 1.

CHAPTER SIX

Adapting to Climate Change: Policies and Actions

ZHUANG GUIYANG AND BO FAN*

1. INTRODUCTION

Over the past century, China has followed a mostly consistent trend of climate change with the world in general. Rising temperatures, faster sea level rise than the global average, frequent extreme weather events and new environmental problems, such as haze and ozone pollution, have posed challenges to mankind. Overall, as climate change has done more harm than good to China,[1] addressing climate change has become a major challenge on China's development path. China participated in international climate negotiations since the 1990s, promoted ecological conservation at the beginning of the twenty-first century, and now leads the world in addressing climate change. China has made commendable

---

* Zhuang Guiyang, Doctor of Economics, Research Fellow of Institute of Urban Environment of Chinese Academy of Social Sciences, Secretary-General of CASS Think Tank for Eco-civilization Studies, Doctoral Supervisor of Graduate School of Chinese Academy of Social Sciences, research fields: low-carbon economy and climate change policy, ecological civilization construction and green development; publications: Low-Carbon Development Blueprint for China's Cities: Integration, Innovation and Application, Low-Carbon Economy: China's Development Path under Climate Change, International Climate System and China. Bo Fan, doctoral candidate at Graduate School of Chinese Academy of Social Sciences, research field: economics of sustainable development.

achievements over the past three decades. However, we must be soberly aware that China is still in the middle stage of urbanization and industrialization, with increasing energy and resource consumptions and demands as well as a huge pressure to reduce emissions; worse still, international climate governance pattern has changed repeatedly, making it difficult to achieve emission reduction goals. Therefore, addressing climate change is a long-term arduous task for both China and the world. To this end, we shall firm our confidence in addressing climate change. Only by learning experience and lesson from previous mitigation and resilience actions and "doing a good job in climate diplomacy," can we meet the new requirements put forward at the Nineteenth CPC National Congress with ease, which are to "take a driving seat in international cooperation to respond to climate change, China has become an important participant, contributor, and torchbearer in the global endeavor for ecological conservation."

2. EVOLUTION AND OVERALL ACHIEVEMENTS OF CHINA'S CLIMATE POLICY

2.1. Evolution of China's climate policy

The *United Nations Framework Convention on Climate Change* (hereinafter referred to as the "UNFCCC"), which was adopted at the United Nations Conference on Environment and Development (UNCED) in 1992, became the world's first legal document on controlling greenhouse gas emissions. China actively participated in the conference and made a great contribution in the conclusion of the UNFCCC. Accordingly, China was listed among Non-Annex I parties, and then has made explorations for addressing climate change, established systems, and deepened and promoted them in the trend of global climate governance.

2.1.1. Starting period of China's effort in addressing climate change (1978–2006)

Since the 1990s, China began relevant research on climate change by improving scientific understanding and initially setting targets for domestic energy conservation and pollution control. The Chinese government established the National Coordination Group on Climate Change in 1990 and renamed it the National Coordination Group on Climate Change Countermeasures in 1998, with the aim to study and review international cooperation and negotiation counterproposals and formulate climate policies. The authority in charge of daily climate work shifted from China Meteorological Administration to the National Development and Planning Commission (now known as the National Development and Reform Commission), suggesting that the Chinese government regarded the climate issue as a development issue.[2] In 1994, China formally adopted *China's Agenda 21* as the guidelines for the implementation of a sustainable development strategy in China, which, however, stressed that economic growth was the primary condition for solving development problems. An average annual energy conservation rate of 5% was put forward in the *Outline of the Ninth Five-Year Plan* for the first time; according to the *Outline of the Tenth Five-Year Plan*, emissions of major pollutants were required to reduce by over 10%, and energy conservation and emission reduction were officially included in economic and social development goals. Since 2002, the Ministry of Science and Technology, China Meteorological Administration, the Chinese Academy of Sciences and other ministries and commissions have jointly made a comprehensive assessment on China's climate change issue, and published three national climate change assessment reports in 2007, 2011 and 2015, laying a scientific foundation for China's climate governance.[3]

2.1.2. Punning period of China's effort in addressing climate change (2007–2014)

Since 2007, China has gradually improved its systems and mechanisms for addressing climate change, ushering into an era of formal institutional establishment for tackling climate change.[4] At that time, China was enjoying a rapid economic growth, with prominent environmental conflicts and sharply increased pressure to reduce emissions. According to the US Energy Information Administration's statistics on carbon dioxide emissions from burning of fossil energies (like coal, oil and natural gas), China's carbon emission (4.052 billion tons) exceeded that of twenty-eight EU countries (3.942 billion tons) in 2003; China's figure (5.912 billion tons) went beyond that of the US (5.602 billion tons) in 2006, making China the world's largest carbon emitter.[5] At this turning point of transition, the report of the Seventeenth CPC National Congress put forward ecological advancement and in-depth implementation of the *Scientific Outlook on Development*, and stressed to "strengthen capacity building for addressing climate change and make new contributions to the protection of global climate." Under the guidance of the *Scientific Outlook on Development*, the State Council set up the National Leading Group on Climate Change, Energy Conservation and Emissions Reduction in 2007; then the latter issued the *China's National Climate Change Program*, which clearly set the goals of "through these measures, the energy consumption per-unit GDP is expected to drop by about 20% by 2010 compared to that of 2005, and carbon dioxide emissions will consequently be reduced." In 2008, the National Development and Reform Commission set up the Department of Climate Change in charge of relevant specific work, including policy formulation, international negotiations, capacity building and carbon market construction;[6] meanwhile, special functional agencies for addressing climate change were established in many provinces and municipalities, marking China's first step in establishing regulations and systems for addressing climate change. The report to the Eighteenth CPC National Congress further lifted ecological conservation to the height of a national strategy in the Five-Sphere Integrated Plan, stressed to "will work with the international community to actively respond to global climate change on the basis of equity and in accordance with the common but differentiated responsibilities and respective capabilities of all countries," and further defined China's determination and basic path to address climate change.

During the Eleventh Five-Year Plan period and the Twelfth Five-Year Plan period, energy consumption, emission control and other constraint indicators were incorporated into socioeconomic development goals, with increased standards and more comprehensive requirements, pushing efforts in addressing climate change from overall planning to concrete implementation. The *Outline of the Eleventh Five-Year Plan* required reducing energy consumption per unit of GDP by 20% and total missions of major pollutants by 10% and increasing the forest coverage to 20%, with fruitful achievements in controlling greenhouse gas emissions. Prior to the United Nations Climate Change Conference in 2009, the Chinese government promised to reduce carbon dioxide emissions per unit of GDP by 40% to 45%, increase the proportion of non-fossil energy in primary energy consumption to about 15%, the forest area by 4000hm^2 and the forest growing stock by 1.3 billion m^3 compared with 2005 by 2020, fully demonstrating China's sincerity in reducing emissions. The *Outline of the Twelfth Five-Year Plan* further put forward the rigid requirements of reducing energy consumption per unit OF GDP by 16% and carbon dioxide emissions by 17% and increasing the proportion of non-fossil energy in primary energy consumption to about 11.4% compared with 2010 by 2015. In 2014, China announced in the *China-US Joint Announcement on Climate Change* that it intends to achieve the peaking of CO2 emissions around 2030 and to make best efforts to peak early and form a mechanism for forcing transition and development through its commitment to absolute emission reduction.

2.1.3. Deepening period of China's effort in addressing climate change (2015 to present)

As China's economy has entered a new normal state, China places more emphasis on improving quality and transforming the driving force in economic development, in an attempt to follow a green and low-carbon development path through supply-side structural reform and free its economic growth from dependence on energy consumption. During this period, efforts shall be made to improve the top-level design for addressing climate change step by step. The *National Plan on Climate Change* (2014–2020) issued by National Development and Reform Commission shall be taken as medium- and long-term guidelines for addressing climate change in China. The *Outline of the Thirteenth Five-Year Plan* had a special chapter on "actively addressing global climate change" and expanded resource and environment-related indicators to 10 items, including reducing energy consumption per unit of GDP by 15%, increasing the proportion of non-fossil energy in primary energy consumption by 3%, and decreasing carbon dioxide emissions per unit of GDP by 18%. The *Work Plan for Controlling Greenhouse Gas Emission during the Thirteenth Five-Year Plan period* required reducing carbon dioxide emissions per unit of GDP by 18% compared with 2015, giving priority to development zones in peaking and strengthening control over non-carbon greenhouse gas emissions. In practice, various pilot projects, such as low-carbon cities, low-carbon parks, sponge cities and climate-adaptive cities, have been emerging nationwide. Besides, carbon trading markets has been promoted from pilot projects to the whole country, in order to explore feasible ways to address climate change.

Facing the new stage, new contradictions and new goals, China has launched a new round of institutional and administrative reforms to improve its governance system and capacity. In 2018, the Fifth Plenary Session of the First Session of the Thirteenth National People's Congress adopted an institutional reform plan of the State Council, which transferred the duties for addressing climate change and reducing emissions to the Ministry of Ecology and Environment, with the aim to strengthen efforts in addressing climate change and providing an institutional guarantee for coordinated planning and promotion of environmental protection and climate governance.

Through its continuous actions to reduce emissions, China has accumulated rich experience in its economic transition and climate governance and become more "confident" in play a leading role in international climate negotiations. Before Paris Climate Conference in 2015, the Chinese government submitted to the secretariat of the UNFCCC a document on nationally determined contributions to climate change entitled *Enhanced Actions on Climate Change: China's Nationally Determined Contribution*, which clearly put forward to achieve the peaking of carbon dioxide emissions around 2030, increase the share of non-fossil fuels in primary energy consumption to around 20 by 2030, lower carbon dioxide emissions per unit of GDP by 60% to 65% from the 2005 level by 2030; and increase the forest stock volume by around 4.5 billion cubic meters on the 2005 level, in an attempt to control both carbon intensity and total emissions. In addition to practicing its promise, China also paid attention to improving its discourse power in international climate negotiations, actively mediated in climate negotiations among different camps, and gave constructive suggestions for emission reduction

2.2. China's achievements in addressing climate change

After years of efforts, China has made significant progress in addressing climate change, becoming the largest carbon emission reduction country.[7] Specifically, it has met all binding targets on schedule, completed a well-established top-level design for addressing climate change, and improved its capacity building constantly.

2.2.1. Decoupling trend between economic growth and carbon emission begins to take shape

In general, since the Twelfth Five-Year Plan period, China has made more efforts in energy conservation and emission reduction and curbed the momentum of continued growth of carbon emissions (see Figure 6-1). After the global financial crisis in 2008, under the guidance of stimulant and expansionary economic policies in China, there was a growth momentum in high-energy consumption and high-emission industries, with the deviated economic growth trajectory from the intensive model and the slight increase in the Tapio decoupling index (the ratio of carbon emission growth to economic growth). Since the beginning of the Twelfth Five-Year Plan period (2011), the growth rate of carbon dioxide emissions was on the decline. In 2015, China's carbon emission from fossil energy's burning dropped by 0.6% for the first time,[8] reversing years of rapid growth in carbon dioxide emissions. The Tapio decoupling index slowly decreased from 0.95 in 2011 to a negative value in 2015 and was only -0.07 in 2016. The economic growth rate and the carbon emission growth rate showed an inverse relationship, indicating the initial decoupling between economic growth and carbon emission. In 2017, the total carbon emissions increased slightly, suggesting arduous tasks in emission reduction. However, the carbon emission intensity still decreased year by year from 34,000 tons/100 million yuan in 2007 to 12,400 tons/100 million yuan in 2017,[9] implying that the low-carbon transition trend of China's economic growth pattern will not change.

Specifically, from 2005 to 2015, China's energy consumption per unit of GDP decreased by 34%; in 2015, carbon dioxide emissions per unit of GDP reduced by 38.6% compared with 2005 and by 21.7% compared with 2010.[10] A total of 4.1 billion tons of carbon dioxide emissions were reduced, exceeding the goals for addressing climate change set forth in the Twelfth Five-Year Plan (see Table 6-1).[11] China witnessed an optimized energy structure and declining coal consumption

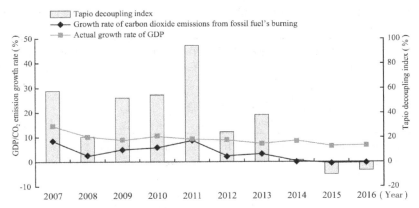

Figure 6-1 Decoupling between China's economic growth and carbon emission
Source: BP Statistical Review of World Energy (2017).

and kept a leading position in nuclear, wind, solar and other renewable energy sectors. The forest coverage increased from 16.55% to 21.66%, making China the country with the largest increase in forest resources in the world. Furthermore, with remarkable progress in air pollution control and improvements in ecological and environmental conditions, China's economic benefits significantly increased.[12]

A number of goals set forth in the Twelfth Five-Year Plan were fulfilled ahead of schedule. During the Twelfth Five-Year Plan period, carbon intensity dropped cumulatively by 20%, exceeding the goal of reducing by 17% in the Twelfth Five-Year Plan; the energy structure was further improved, as evidenced by the proportion of non-fossil energy in primary energy consumption of 12% in 2015, exceeding the goal of accounting for 11.4% set forth in the Twelfth Five-Year Plan; the forest growing stock increased by about 2.68 billion m³ compared with 2005, fulfilling the goal of increasing the forest growing stock by 2020 ahead of schedule.

At present, there is a good progress of achieving the goals set forth in the *Thirteenth Five-Year Plan*. In 2016, China's carbon intensity reduced by 6.6%, down by 42% from 2005, exceeding the goal of a 40% to 45% reduction by 2020.[13] In addition to the medium-high growth rate in GDP and the increasing economic aggregate, the proportion of coal continued to decline, and the proportion of non-fossil energy rose rapidly. In particular, China ranked first in the world in terms of installed hydropower capacity, nuclear power capacity under construction, solar heating area, installed wind power capacity and artificial afforestation area.[14]

Table 6-1 Energy conservation and emission reduction constraint goals in five-year plans and their fulfillments

	The Eleventh Five-Year Plan (2006–2010)		The Twelfth Five-Year Plan (2011–2015)		The Thirteenth Five-Year Plan (2016–2020)	
	Planning goals	Fulfillments	Planning goals	Fulfillments	Planning goals	Fulfillments
Reduced energy consumption per unit of GDP (%)	20	19.2	16	Cumulative average annual growth rate at 18.2	15	—
Reduced carbon dioxide emissions per unit of GDP (%)	Down by 17 year on year	16	17	Cumulative average annual growth rate at 20	18	
Total carbon dioxide emissions (%)	—	—	—	—	Peak in 2030	—
Proportion of non-fossil energy in primary energy consumption (%)	10	8.6	11.4	12	15	—
Forest coverage (%)	20	20.36	21.66	21.66	23.04	—
Forest stock volume (billion m^3)	137.55	137.21	143	151.37	165	—

Source: *The Outline of the Eleventh Five-Year Plan*, the *Outline of the Twelfth Five-Year Plan*, the *Outline of the Thirteenth Five-Year Plan*.

2.2.2. Multilevel management systems and policy systems are established to address climate change

In terms of organizational establishment, China has experienced functional changes in China Meteorological Administration, the National Development and Reform Commission and the Ministry of Ecology and Environment. To address climate change, administrative systems and work mechanisms have taken shape, with unified and coordinated management of the Ministry of Ecology and

Environment of the State Council, work division and separate responsibilities of relevant departments and local departments, powerful support from think-tank agencies and extensive participation of the whole society. As the assessment of responsibilities for fulfilling emission reduction goals has been strengthened, a greenhouse gas emission accountability system has been established in accordance with the *Work Plan for Controlling Greenhouse Gas Emissions during the Thirteenth Five-Year Plan period*, requiring to strengthen the assessment and examination on provincial people's governments for their fulfillment in controlling greenhouse gas emissions.

In terms of policy system, China has not only approved the *UNFCCC*, the *Kyoto Protocol* and the *Paris Agreement* but also formulated various mitigation and resilience action plans based on its national conditions. According to the requirements of the UNFCCC, China submitted the *Initial National Communication on Climate Change of the People's Republic of China* and the *Second National Communication on Climate Change of the People's Republic of China* in 2004 and 2016, respectively, so as to make clear greenhouse gas inventories and the impacts of climate change and define the direction of addressing climate change. In 2007, *China's National Plan for Addressing Climate Change* was released as basic guidelines for all climate policies. Since 2008, the Chinese government has successively published annual reports on *China's Policies and Actions for Addressing Climate Change*, clarifying the process and basic path of China's efforts in addressing climate change. The *National Strategy for Climate Change Adaptation* (2013–2020) and the *National Plan on Climate Change* (2014–2020) were issued to incorporate "mitigation" and "adaptation" into China's policy system for addressing climate change. In addition, the *Law on Renewable Energy*, the *Law on Promoting Clean Production*, the *Law on the Prevention and Control of Environment Pollution Caused by Solid Wastes*, the *Circular Economy Promotion Law* and other special legislations were issued, as well as the *Work Plan for Controlling Greenhouse Gas Emissions during the Twelfth Five-Year Plan period*, the *Work Plan for Controlling Greenhouse Gas Emissions during the Thirteenth Five-Year Plan period*, the *Twelfth Five-Year Plan for Energy Conservation and Emission Reduction*, the *Thirteenth Five-Year Plan for Energy Conservation and Emission Reduction* and other detailed action plans were approved, building the institutional foundation addressing climate change along with national programs. Local legislations on addressing climate change in Qinghai, Heilongjiang and Sichuan have been launched and even entered the stage of soliciting opinions, providing a practical foundation for studying and making laws on addressing climate change at the national level.[15]

2.2.3. Capacity building for addressing climate change is strengthened constantly

In terms of funding, China established a stable channel of government funds, including funds from the national science and technology program to address climate change, special funds for energy conservation and emission reduction as well as investment funds for low-carbon industries. In the meantime, it attracted social funds to address climate change and raised funds through multiple channels. From 2010 to 2014, the government spent 821.069 billion yuan in support of actions relating to mitigation and resilience to climate change.[16] To achieve the goal of nationally determined contributions, from 2005 to 2015, China had invested RMB10.4 trillion yuan and from 2016 to 2030, it will invest another 30 trillion yuan.[17] "South-South Cooperation Assistance Fund" is a financial mechanism independently established by China to work with developing countries on climate change, in order to help developing countries to raise funds from Green Climate Fund and other financial institutions or international institutions.[18] As a developing country, China is eligible as a recipient county of the climate finance support under the UNFCCC and has got involved in programs for addressing climate change supported by Global Environment Facility and other organizations.

In the aspect of science and technology, China independently compiled greenhouse gas inventories and launched special technical actions for addressing climate change. According to the requirements of the UNFCCC, China established a national greenhouse gas inventory database to support the inventory compilation and data management under the leadership of the National Development and Reform Commission in cooperation with domestic research institutions and universities, and has compiled greenhouse gas inventories for 1994, 2005 and 2012.[19] Since the implementation of Eighth Five-Year Plan, China has organized a series of scientific and technological projects relating to climate change through National Programs for Science and Technology Development, National High-tech R&D Program of China and National Basic Research Priorities Program, with the aim to participate in international scientific and technological cooperation on global climate change and improve basic research on climate change.[20] Furthermore, it issued *China's Scientific & Technological Actions on Climate Change*, *Study on Climate Change Adaptation Strategy of Science and Technology Development in China*, and *Special Program Plan for Scientific and Technological Innovation to Address Climate Change during the Thirteenth Five-Year Plan Period* one after another, forming the scientific and technological action guidelines for addressing climate change.

In talent building, after years of efforts, China has formed a cross-disciplinary and cross-domain expert team engaged in basic research and applied research in the field of climate change, made a number of pioneering research achievements, and completed many state-level scientific research bases and a large observation network system, including the national climate monitoring network, in a bid to provide important scientific and technological support for China's efforts in addressing climate change.[21] Relevant departments have organized climate change addressing capacity training to enhance their work capacity to address climate change.[22] The Ministry of Education has encouraged secondary and higher education institutions to independently open majors relating to climate change and cultivate professionals in the field of climate change.[23]

3. CLIMATE CHANGE MITIGATION ACTIONS

"Mitigation" and "adaptation" are the dual pillars for addressing climate change. Over the past decade, China has made remarkable achievements in its vigorous policies to shut down outdated production capacity and improve energy efficiency, making great contributions to emission reduction in the production sector. In recent years, as widespread green buildings, green transport and green products have led to a reform of green consumption pattern, carbon emission reduction in life will be a key task in mitigation actions in the future. As resilience to climate change emphasizes enhanced resilience capacity, flexible response and management of climate risks are particularly important for developing countries that give consideration to development demands and climate risks in a coordinated manner.

3.1. Building a modern green economy through overcapacity cut and restructuring

Since the late 1980s, the Chinese government has paid more attention to the transformation of the economic growth pattern and the adjustment of the economic structure,[24] and promoted the upgrading of the industrial structure by shutting down outdated production capacity and vigorously developing emerging industries.

Outdated production capacity has been shut down to reduce carbon emissions at the source of production. In the twenty-first century, China has bid farewell to the era of shortage economy and turned to the "buyer's market." In the dilemma of "low-end overcapacity and high-end capacity shortage," the *Outline of the Eleventh Five-Year Plan* focused on integrating coal capacity and resolving overcapacity in steel, cement and other industries, and took overcapacity cut as an important way to change the economic development pattern. During the Twelfth Five-Year Plan period, more efforts were made to cut overcapacity. According to the *Outdated Production Capacity Elimination Goals in Key Industrial Sectors during the Twelfth Five-Year Plan Period*, outdated production capacity in nineteen key industries were determined to be the elimination goals. Since 2013, as China's economy has entered a new normal state, it is more necessary to push supply-side structural reform by means of overcapacity cut, in a bid to "resolve, transfer, integrate and eliminate" overcapacity.[25] In 2016, the *National Development and Reform Commission Issued the Circular on Reducing and Substituting Coal Consumption in 2016*, which proposed to strictly control steel, coal, cement clinker and other industries with overcapacity as well as coal and power industries with potential overcapacity risks, control total coal consumption, and continue to promote the transformation and upgrading of traditional industries.

Transformation and upgrading of industrial structures have been promoted to reduce carbon emission intensity. Since the Eleventh Five-Year Plan period, China has taken circular economy as a key development strategy, and then supported a number of circular economy demonstration and pilot projects and circular economy industrial parks.[26] Since the Twelfth Five-Year Plan period, China's service sector and emerging industries have developed rapidly. According to the *Twelfth Five-Year Plan for Development of Strategic Emerging Industries*, energy conservation and environmental protection, next-generation information technology, biology, high-end equipment manufacturing, new energy, new materials and new-energy vehicles were listed as key development directions of strategic emerging industries. The *Outline of the Thirteenth Five-Year Plan* put forward to promote new-energy vehicles, new energy, energy conservation and environmental protection, and other green and low-carbon industries and systematically advance the "Made in China 2025" and "Internet +" actions, so as to promote the inner structure of industry to move toward medium-high end. In 2012, the proportion of added value of the tertiary industry in China's GDP was the almost same as that of the secondary industry for the first time. In 2015, the proportion reached 50.2%.[27] New driving forces, new industries and new forms of business have grown rapidly. From 2013 to 2017, high-tech manufacturing grew at an average annual rate of 11.7%, and the proportion of the service sector increased from 45.3% to 51.6%,[28] becoming the main driving force of economic growth. In brief, a low-energy, low-emission industrial structure will be a key factor for reducing carbon emission intensity.

3.2. Establishing a clean energy system through energy conservation and emission reduction

Energy conservation management has been strengthened. The construction of a resource-intensive, environment-friendly society is a major strategic task in the *Outline of the Eleventh Five-Year Plan*. In 2007, the national steering group on climate change, energy conservation and emission reduction were established to give high priority to energy conservation and emission reduction, issue the

Comprehensive Work Plan for Energy Conservation and Emission Reduction, make comprehensive arrangements for energy conservation and emission reduction, and set up an accountability for fulfilling energy conservation and emission reduction goals. The *Twelfth Five-Year Comprehensive Work Plan for Energy Conservation and Emission Reduction* proposed energy conservation and transformation projects, energy-saving technology industrialization demonstration projects, subsidizing the production of energy-efficient products, energy performance contracting promotion projects and energy-saving capacity building projects. China has also adopted such measures as the "Energy Saving and Low Carbon Action Implementation Program in Top-10000 Enterprises," management of energy-saving standard logos, promotion of energy-saving technologies and products, action for "100 key units with the highest energy consumption, 1,000 key units with a higher energy consumption and 10,000 key energy-consuming units," energy conservation assessment for public institutions, and strict control of energy consumption in construction, transportation and other key sectors. At present, China's total energy consumption is under effective control, and energy consumption intensity is on the decline. In 2016, energy consumption per unit of GDP fell by 5% year on year, which was equivalent to 500 million tons less carbon dioxide emissions.[29]

Efforts have been made to promote clean utilization of fossil energy and vigorously develop renewable energy. The *Renewable Energy Law* and a series of allocation policies and measures were issued to provide a stable support platform for leapfrog development of new and renewable energy in China during the Eleventh Five-Year Plan period.[30] Since the Twelfth Five-Year Plan period, vigorous efforts have been made to promote clean utilization of fossil energy, and upgrade coal-fired power plants. In 2015, the average net coal consumption rate of coal-fired power generation units with a capacity of more than 6,000kW in China was about 315g standard coal/KWH, with a cumulative reduction of 18g standard coal/KWH within five years.[31] The resource tax reform was promoted in an all-round way, as evidenced by the fact that the coal resource tax was transformed into ad valorem resource tax, which forced enterprises to change their production mode to solve negative externalities in resource development. The "coal-to-gas" project was advanced to substitute coal consumption. Meanwhile, priority was given to development of renewable energy through such policies as compensating renewable energy costs and deepening reform of the electric power system. As a result of the energy reform, the proportion of non-fossil energy consumption in China's total energy consumption rose from 6.9% in 2000 to 13.3% in 2016,[32] indicating significant improvements in energy structure.

3.3. Developing a green consumption pattern and lifestyle

With the increase of the proportion of tertiary industry and the per capita income level, the proportion of carbon emission from household life has risen rapidly. After global financial crisis in 2008, drivers of China's economic growth have gradually shifted from investments to consumptions. In 2017, final consumption contributed up to 58.8% to economic growth.[33] According to the Report to the Nineteenth CPC National Congress, "The fundamental role of consumptions in economic development shall be strengthened." Carbon emission reduction in consumption and life will become the emphasis of emission reduction in the future.

The *Outline of the Tenth Five-Year Plan* clearly put forward to "raise overall environmental awareness and promote green consumption." In 1999, the National Economic and Trade Commission, the National Environmental Protection Administration, and other ministries and commissions jointly proposed and implemented the "three green" program for "advocating green consumption, cultivating green market and opening up green channels," with the aim to deepen the concept of green consumption into people's life. From then on, energy-saving product certification, green food certification, government procurement of energy-saving products, plastic limit and other measures have been launched, so as to push energy conservation and emission reduction in the consumption field based on systems; progressive pricing for household electricity and water consumption has been implemented to fully reflect the scarcity of resources and energy and promote recycling and economical utilization. In the field of urban construction, the Ministry of Housing and Urban-Rural Development issued the first edition of the *Assessment Standard for Green Building* in 2006, indicating the start of development and application of green buildings. Such measures as new-energy vehicles, public transport system and green logistics chain were adopted to strictly control carbon emission in the transportation sector. After the Nineteenth CPC National Congress, the Ministry of Transport proposed to initially build an eco-friendly, clean, low-carbon, intensive and efficient green transport system with a science-based layout by 2020.[34] Green consumption and green production will form a two-way constraint mechanism to reduce carbon emission in production and life.

3.4. Increasing carbon sink

Since the 1980s, the Chinese government has continuously increased investments in afforestation projects and mobilized citizens of the right age to participate in nationwide voluntary tree planting. It's estimated that from 1980 to 2005, 3.06 billion tons of carbon dioxide had been absorbed in China's afforestation activities, 1.62 billion tons of carbon dioxide had been absorbed by forest management and 430 million tons of carbon dioxide emissions from deforestation had been reduced, enhancing the greenhouse gas sink capacity effectively.[35] According to the *Outline of the Twelfth Five-Year Plan*, the forest growth indicator was included in the assessment as an obligatory target, and the completion of increasing forest carbon sink was included in the assessment of responsibility for reducing carbon dioxide emissions per unit of GDP.[36] Besides, in Beijing-Tianjin-Hebei ecological forestry construction, three norths and Yangtze River Basin shelterbelts and other key projects, a total of 460 million mu of forests were planted in the past five years.[37] On the basis of the latest report of the UN Food and Agriculture Organization in 2015, "China recorded the largest net increase of forest area in the world from 2010 to 2015, setting a good example for the world."[38] The *Outline of the Thirteenth Five-Year Plan* called for a complete cease of commercial clear-cutting of natural forests and greater efforts in protecting forest ecosystems.

In terms of forest carbon sink, China has promoted forest management, established a comprehensive forest carbon sink measurement and monitoring system, strengthened management of technical standards for forest carbon sinks, compiled an inventory of forest carbon sinks, and further promoted the forest carbon sink trading system. Currently, there are four main sponsors of forest carbon sink trading projects in China: Clean Development Mechanism (CDM), China Certified Emission Reduction (CCER), Verified Carbon Standard (VCS) and China Green Carbon Foundation (CGCF). In addition, China has also been active in wetland protection and restoration, ecological protection and construction for grasslands and development of marine carbon sink, in a bid to increase carbon sinks through multiple channels.

4. CLIMATE CHANGE RESILIENCE ACTIONS

The concept of climate change resilience was first put forward in *China's Agenda 21*. In 2007, the *Bali Action Plan*, which was adopted at the Thirteenth Conference of the Parties to the UNFCCC, placed equal importance on climate change resilience and mitigation. In the same year, China issued the *National Plan on Addressing Climate Change* to systematically describe various resilience tasks. The *National Strategy for Climate Change Resilience*, which was published in 2013, incorporated climate change resilience into the government's economic and social development plans, and divided China's key areas into three resilience zones: urbanization, agricultural development and ecological security, defining China's strategic pattern for climate change resilience.

4.1. Construction of climate-resilient cities

Cities are the gathering places of population and industries, with a higher risk exposure. In response to rainfall flood management, the Ministry of Housing and Urban-Rural Development set up a total of thirty sponge cities in two batches in 2015 and 2016.[39] It aimed to stress that cities can be as resilient as sponges in environmental changes and natural disasters and explore a low-impact development way based on China's national conditions. According to the *Guiding Opinions on Promoting Construction of Sponge Cities* issued by the General Office of the State Council in 2015, the control rate of annual total runoff of rainwater and flood regulation and storage shall be no less than 70% by 2020, and that in pilot area shall be no less than 75%, indicating the initial establishment of urban sponge bodies.

In response to "road zippers" and "air webs" in urban construction, the General Office of the State Council issued the *Guiding Opinions on Promoting Construction of Urban Underground Integrated Pipe Galleries* in 2015, requiring to build urban underground integrated pipe galleries covering urban water supply, drainage, gas, heating, electricity, communication pipelines and ancillary facilities. It planned to launch the underground integrated pipe gallery pilot project in thirty-six large and medium-sized cities in about three years, with the aim to save underground space and resources and improve urban services in an all-round way.

In view of the contradictions between urban expansion and ecosystem destruction, the Ministry of Housing and Urban-Rural Development appointed Sanya as the first pilot city of "urban repair and ecological restoration" in 2015, with remarkable achievements. In 2017, the *Ministry of Housing and Urban-Rural Development Issued the Guiding Opinions on Strengthening Urban Repair Work for Ecological Restoration*, and set up fifty-seven pilot cities,[40] with the aim of treating "urban diseases," transforming the urban development pattern, and restoring urban ecosystems comprehensively.

Compared with the above three types of pilot cities, climate-resilient pilot cities have broader connotations, with an improved urban resiliency in all aspects, like infrastructure, ecosystem and management system. In 2016, the National Development and Reform Commission and the Ministry of Housing and Urban-Rural Development issued the *Urban Climate Change Resilience Action Plan*, which made deployments for urban infrastructure construction, urban building resiliency, urban ecological greening function, urban water security and integrated disaster risk management system, and proposed to build thirty climate-resilient pilot cities by 2020 according to climate risk, urban size and urban function of different cities. At the beginning of 2017, twenty-eight climate-adaptive pilot urban areas were launched,[41] taking the lead in carrying out adaptation actions.

4.2. Building agricultural resiliency

China's agricultural production is highly labor-intensive and constrained by land and water resources. Uncertain impacts of climate change on agriculture are mainly reflected in its direct effect on biological mechanisms of crops and its indirect effect on agricultural production through its impacts on water.[42] To this end, China has built agricultural infrastructures, improved construction of irrigation and water conservancy facilities, promoted development of key irrigation and water conservancy counties, and advanced water-saving irrigation projects. Specifically, it has carried out "water conservation and grain increase action" in northeast China and built eleven high-standard water-saving agricultural demonstration zones in eleven provinces, autonomous regions and municipalities in north, northwest and southwest China. In terms of crop production, efforts have been made to promote conservation farming, launch the "action of zero increase in use of pesticides and chemical fertilizers by 2020," adjust the agricultural production structure, and develop multiple sustainable livelihood industries, including energy forestry, agricultural product processing and ecotourism. In 2013, the Ministry of Agriculture and the Global Environment Facility jointly carried out a five-year climate-smart agriculture pilot and demonstration project in major grain-producing areas, in order to enhance climate change resilience of crop production.

4.3. Ecosystem protection

Unified management and protection for water resources have been strengthened. By formulating integrated plans for water resources and river basins, China has initially established the river basin flood control and disaster mitigation system, the rational water resource allocation system and the water resource protection system.[43] Water conservancy projects have been strengthened, including flood control projects, water supply projects and interconnected river system network projects. The assessment of water resource management system has been implemented, covering all thirty-one provinces, autonomous regions and municipalities. During the Thirteenth Five-Year Plan period, efforts were made to conduct two-way control for total water consumption and intensity, build a water-saving society, and carry out ten water-saving action plans nationwide, including water conservation and production increase in agriculture, water conservation and efficiency in industry, and water conservation and consumption reduction in urban areas.

Terrestrial ecosystem protection has been strengthened. Water and soil erosion control has been intensified to reverse the desertification trend and shrink desertified land area year by year; in particular, China's desertified land area has decreased by an average of 1,980 square kilometers per year in the past five years.[44] In addition, efforts have also been made to carry out projects for returning farmland to forests and returning grazing land to grassland, implement grassland grazing prohibition and forage-livestock balance systems, establish a national park system pilot program[45] and improve the protected natural area system.

Marine ecosystem system has been protected. Since 2008, China has established a work system for addressing climate change in marine areas and issued a planning system for addressing climate change in marine areas.[46] Furthermore, the marine disaster emergency plan system and the response mechanism have been strengthened. Marine ecosystem protection and restoration have been strengthened by establishing typical marine ecosystem restoration demonstration zones and implementing the "blue" bay remediation and restoration action and the "ecological islands and reefs" project, and pollution in offshore areas has been controlled to improve the marine ecosystem environment. Coastal zone management has been strengthened by enhancing protection standards for coastal cities and major engineering facilities. By July 2017, about 260 various types of marine protected areas had been established at all levels.[47]

4.4. Construction of disaster prevention and mitigation systems

Since 2009, China has formulated the *National Meteorological Disaster Prevention Plan* (2009–2020), the *National Comprehensive Disaster Prevention and Mitigation Plan* (2016–2020), and the *Administrative Measures for National Emergency Warning Information Release System*, forming a basic institutional framework for climate disaster prevention and mitigation. In 2014, warning information was automatically connected at the national level, and national, provincial, municipal and county-level meteorological disaster risk warning systems were established.[48] During the Twelfth Five-Year Plan period, the National Disaster Mitigation Commission and the Ministry of Civil Affairs launched 158 national emergency responses to various natural disasters in total.[49] During the Thirteenth Five-Year Plan period, efforts were made to get social forces involved in the disaster relief work mechanism, and explore and introduce market means to promote climate risk and disaster management. For example, it was planned to expand the agricultural disaster insurance pilot program and the types of insurance to support the insurance business in agriculture and forestry sectors.[50]

5. LOW-CARBON PILOT AND DEMONSTRATION SYSTEM AND CARBON MARKET DEVELOPMENT

Since the Twelfth Five-Year Plan, China has deepened low-carbon pilot programs at levels of province, municipality, town, industrial park and community.[51]

The pilot areas have served as pioneers in terms of planning concepts, management pattern, systems and mechanisms, and evaluation standards, as well as "test fields" for climate policy. In 2017, carbon emission trading market was promoted from the pilot areas to nationwide, giving play to the market mechanism function, with a great significance to industrial restructuring and transformation of economic development pattern.

5.1. Construction of low-carbon pilot and demonstration system

The National Development and Reform Commission appointed three batch of low-carbon pilot provinces and municipalities and low-carbon pilot cities in 2010,[52] 2012[53] and 2017[54]. In the selection process of low-carbon pilot cities, such characteristics as regional endowment, economic pattern and development stage were taken into full consideration; the pilot program was gradually extended from the provincial level to the municipal level, covering all of eastern, central and western areas, and then rolled out nationwide from point to area. In accordance with the requirements of the national deployment and implementation plan, all pilot areas formulated low-carbon development plans, established greenhouse gas emission statistics and management systems, developed a low-carbon production pattern and lifestyle, set up low-carbon evaluation and examination mechanisms, and improved the low-carbon development management capacity.

After seven years of efforts, low-carbon pilot cities have grown into the main force in China's low-carbon development. Currently, all pilot provinces and municipalities have completed the *Low-Carbon Development Plan* or the *Plan for Addressing Climate Change*, set clear targets for peaking, initially set up a greenhouse gas inventory compilation system, and adopted corresponding policies and measures in organizational management, statistical accounting and evaluation systems, target-oriented responsibility system, industrial policies, fiscal and tax policies, special low-carbon development funds, government procurement and carbon emission trading pilot program. According to the greenhouse gas emission control target responsibility assessment and examination in 2012 organized by National Development and Reform Commission, the carbon intensity of the ten provinces and municipalities included in the pilot program decreased by 9.2% on average in 2012 compared with that in 2010, which was higher than the 6.6% overall reduction throughout the country,[55] with significant achievements on carbon emission reduction. All pilot cities made innovations in low-carbon development policies. For example, Shenzhen's government formulated China's first local special regulation for standardizing carbon emission and carbon trading – the *Administrative Regulations on Carbon Emission in Shenzhen Special Economic Zone*, and launched China's first carbon bond, carbon fund, private carbon fund, Internet carbon financial product "Peiebao" and other carbon financial service products; local governments of Qingdao, Hangzhou and Zhenjiang developed carbon emission information technology management platforms. These excellent experiences have laid a foundation for the promotion of low-carbon development pattern nationwide.

Moreover, the Ministry of Housing and Urban-Rural Development proposed to build low-carbon ecosystem pilot cities (towns) in newly built cities (towns) and new areas of existing cities in 2011. In 2015, the National Development and Reform Commission issued the Guidelines for Construction of Low-carbon Pilot Communities, which made it clear that a number of distinctive low-carbon pilot communities would be created based on their various regional conditions and different development levels. The *Proposal for the Thirteenth Five-Year Plan* put forward for the first time to implement near-zero carbon emission zone demonstration projects. These low-carbon communities and small towns have been complementary with cities, forming an all-round, multilevel low-carbon pilot and demonstration system.

5.2. Construction of carbon emission permit trading market

There are three carbon trading mechanisms based on Kyoto Protocol: clean development mechanism (CDM), joint implementation (JI) and emissions Trading (ET). The *Measures for Operation and Management of Clean Development Mechanism Projects*, which was carried out in China since 2005, marked the beginning of China's participation in the international carbon market as a developing country.

In 2010, the *Decision of the State Council on Accelerating Cultivation and Development of Strategic Emerging Industries* clearly proposed to "establish and improve the major pollutant and carbon emissions trading system." In 2011, the National Development and Reform Commission announced the establishment of seven carbon emission permit trading pilot provinces and municipalities, namely Beijing, Shanghai, Tianjin, Guangdong, Shenzhen, Chongqing and Hubei. All of the pilot provinces and cities officially came into operation in 2013 and 2014, and issued corresponding policies and regulations to explore how to use the carbon market as a market means to control greenhouse gases. China Certified Emission Reduction (CCER), CDM and other carbon offset projects and mechanisms carried out in China since 2012 could appropriately reduce the performance costs for enterprises and bring some benefits. Due to differences in energy consumption and carbon emission, economic development level and local government supervision, various pilot areas varied greatly in market performance, but with a higher performance rate in general.[56] As of November 2017, the seven pilot areas recorded a cumulative quota transaction volume of more than 200 million tons of carbon dioxide equivalent, with a transaction amount of more than 4.6 billion yuan; besides, both total carbon emissions and intensity reduced within the pilot areas, which played a role in controlling greenhouse gas emissions.[57] Compared with other pilot areas, Shenzhen was better in market construction, such as in market resource allocation capacity and environmental constraint.[58]

At the end of 2017, the national carbon emission trading system was officially launched in the principle of "implementing from easy to difficult step by step," with the power generation industry as the starting point. In the future, it would extend to chemical, petrochemical, steel, non-ferrous metal, building material, paper-making and aviation industries step by step. According to the *National Carbon Emission Permit Trading Market Construction Plan (Power Generation Industry)* issued by the National Development and Reform Commission, the national carbon market will be developed in infrastructure construction period, simulation operation period, and deepening and improving period, and advanced step by step. Currently, the national carbon market follows the institutional arrangement of "free allocation of carbon emission quota based on actual output of enterprises," and will force to optimize the electric power structure through carbon constraint, and bring opportunities for clean production and green development, so as to realize "carbon emission with costs and carbon reduction for benefits."

6. GROWING FROM A PARTICIPANT IN, A CONTRIBUTOR TO, AND LEADER OF GLOBAL CLIMATE GOVERNANCE

Based on its national conditions, China has taken an active part in international climate negotiations organized by the United Nations, and increased its discourse power in multilateral governance mechanisms, growing from a "participant" in global climate governance to a "leader and contributor." In addition, China has developed international cooperation by building exchange platforms, providing financial and technological assistance, and jointly building green projects, in a bid to form a road of climate diplomacy with distinctive Chinese characteristics and make outstanding contributions to addressing climate change and safeguarding global ecological security.

6.1. China's presence in global climate negotiation process

6.1.1. Ecological and environmental issues were attached importance with global climate governance actions (1979–2006)

During the period, China paid increasing attention to ecological and environmental issues, actively participated in climate conferences, and negotiated and signed international climate agreements. In 1979, when the 1st World Climate Conference (WCC) was held in Geneva, Switzerland, climate change became a topic of common concern for the international community for the first time. The Chinese government assigned a delegation to WCC and put forward to link climate with social and economic development[59] In 1988, the Intergovernmental Panel on Climate Change (IPCC) under the United Nations was established to assess climate change and its impacts. It published five assessment reports in 1990, 1995, 2001, 2007 and 2014, providing scientific basis for formulating climate policies and promoting the process of international climate negotiations. As China was a contributor to the previous assessment activities, the number of Chinese participants was on the rise, with a growing degree of influence.[60] In 1992, the UNFCCC was approved at the UNCED, which became the world's first international convention for comprehensively controlling greenhouse gas emissions and defined the cooperation principle of "common but differentiated responsibilities." Since 1995, the Conference of the Parties to the UNFCCC has been held almost every year, and China has been an active participant. In 1997, the *Kyoto Protocol* was adopted at the 3rd Conference of the Parties to the UNFCCC and came into force in 2005, which set the greenhouse gas emission reduction task from 2008 to 2012. As a developing country, China didn't undertake the emission reduction task for the time being but faced the risk of taking over high-polluting, energy-energy-consuming industries from developed countries. China officially approved the UNFCCC in 1993 and the *Kyoto Protocol* in 2002. Together with other developing countries, China will supervise and urge developed countries to shoulder their historical responsibilities and fulfill their emission reduction obligations.

6.1.2. An active voice was heard at climate conference to form the "Tripod" pattern (2007–2014)

With the rise of emerging countries, climate change negotiations, a "tripod" pattern has taken place in climate change negotiations, involving the EU, the Umbrella Group,[61] the Group of 77 and China. In 2007, the Thirteenth Conference of the Parties to the UNFCCC formulated the "Bali Roadmap," made arrangements for global response to climate change upon expiration of the first commitment period of the Kyoto Protocol by the end of 2009, and discussed the emission reduction obligations of developed countries and the future emission reduction actions of developing countries. The Chinese delegation put forward separate advance of four wheels, namely "mitigation, resilience, technology and funds," and stressed the importance of "technology and funds" in helping developing countries address climate change. This proposal was incorporated into the Bali Action Plan.[62] In 2009, the 15th Conference of the Parties to the UNFCCC was held in Copenhagen to discuss the subsequent scheme after expiration of the first commitment period of the Kyoto Protocol (2012–2020), and the US and

China, the two largest greenhouse gas emitters, became the focus during the negotiation. Before Copenhagen Climate Change Conference, China issued the Implementation of the *Bali Roadmap—China's Position on the Copenhagen Climate Change Conference*, in order to offer the emission reduction goals and fully demonstrate its sincerity in promoting climate governance. However, the parties didn't reach an agreement in the principle of "common but differentiated responsibilities"; with more serious disputes than ever, they failed to reach a legally binding agreement. Despite the disappointing result at the Copenhagen Climate Change Conference, China tried its best to continue the implementation of the UNFCCC, took a constructive part in negotiations in the new process of "Durban Platform," consolidated its strategic basis in negotiations through "BASIC countries," "like-minded developing countries" and other mechanisms, effectively guaranteed the smooth advance of international climate negotiations and safeguarded the legitimate interests of developing countries.

6.1.3. Efforts were made to elevate discourse power and lead global climate governance (2015–present)

Since 2015, China has fulfilled its commitment to emission reduction, expressed its views and fought for its rights and interests, becoming a leader in the global governance system gradually. At the Paris Climate Change Conference in 2015, nearly 200 parties reached the Paris Agreement, which made arrangements for global response to climate change after 2020, set the goal of "keeping global climate warming below 2°C," and determined that countries will participate in global actions on addressing climate change by means of nationally determined contributions after 2020. Before the conference, China submitted a document on nationally determined contributions to the United Nations Secretariat, held negotiations with India, Brazil, the EU, the US and other countries and regions, and published a series of joint statements, accumulating valuable experience for reaching consensus at the Paris Climate Change Conference.[63] Chinese President Xi Jinping delivered an important speech at the opening ceremony of the Paris Climate Change Conference, stressing, "the Paris Agreement should help meet the goals of the UNFCCC and chart the course for green development. The Paris Agreement should help galvanize global efforts and encourage broad participation. The Paris Agreement should help increase input of resources to ensure actions on climate change. The Paris Agreement should accommodate the national conditions of various countries and lay emphasis on practical results."[64] It charted the course for the key issues in the process of negotiations and expressed China's determination to do its utmost to address climate change and promote green development. After the conference, China took the lead in signing the Paris Agreement, fully demonstrating its sense of responsibility as a major power.

However, since US President Donald Trump announced the US withdrawal from the Paris Agreement in 2017, which casted a shadow over global climate governance, the international community has increasingly called for China to act as a leader. According to the Report of the Nineteenth CPC National Congress, "Over the past five years, China has made remarkable achievements ecological conservation…(China) guided international cooperation on addressing climate change, and became an important participant, contributor and leader in global ecological conservation." This is not only an objective evaluation and basic requirement for China's participation in global climate governance but also a strategic response to international expectations. China's role as a "leader" in global climate governance reflects the inevitable trend for China to follow the evolution of global climate governance pattern as well as an objective choice made by China based on its own national conditions. Chinese President Xi Jinping stressed at the Davos summit in Switzerland in 2017, "The Paris Agreement is consistent with the general direction of global development, and we should adhere to its hard-won achievements together, without abandonment," and "the awareness of the community with a shared future for mankind shall be created to promote global development jointly."[65] This explicitly expressed China's adherence to the multilateralism and the Paris Agreement. In the future, China will assume the responsibility as a major power and contribute Chinese wisdom and Chinese approach to the world with concrete and effective actions.

6.2. Active promotion of international exchanges and cooperation

In addition to leading international climate negotiations, China has also actively deepened cooperation with developed countries, developing countries and NGOs, established dialogue mechanisms, spread the concept of green development, shared governance experience, and built a community with a shared future for mankind hand in hand with the international community, in a bid to jointly address climate change and safeguard global ecological security.

6.2.1. South-South climate cooperation

In the past, South-South cooperation focused on the theme of poverty reduction, with the aim to promote economic and trade exchanges. In the context of global climate change crisis, climate governance has increasingly become a major topic of South-South cooperation. Chinese President Xi Jinping announced at the opening ceremony of the Paris Climate Change Conference "to support developing countries, especially the least developed countries, landlocked developing countries and small island developing states, in confronting the challenge of climate change. In a show of greater support, China announced in September the establishment of an RMB20 billion South-South Climate Cooperation Fund. Next year, China will launch cooperation projects to set up 10 pilot low-carbon industrial parks and start 100 mitigation and adaptation programs in other developing countries and provide them with 1,000 training opportunities on climate change. China will continue promoting international cooperation in such areas as clean energy, disaster prevention and mitigation, ecological protection, climate-smart agriculture, and low-carbon and smart cities. China will also help other developing countries to increase their financing capacity..."[66] Since 2011, with the support of the central budget, the National Development and Reform Commission had carried out South-South cooperation on climate change by donating energy-saving and low-carbon products free of charge and holding climate change trainings. By the end of 2015, the National Development and Reform Commission had signed 22 memorandums of understanding on donating materials for addressing climate change with 20 developing countries, and donated more than 1.2 million LED lamps, 9,000 LED street lamps, 20,000 energy-saving air conditioners and 8,000 solar photovoltaic power generation systems in total; besides, it had held 11 trainings on climate change response and green and low-carbon development for more than 500 officials and technicians in climate change from 58 other developing countries.[67] Meanwhile, China also carried out South-South cooperation on climate change under the framework of the UN, including cooperation with the UN Food and Agriculture Organization, the UN Development Programme and the UN Environment Programme. South-South cooperation on climate change serves as an important platform for China to shoulder its international responsibilities and play a leading role and will be a glorious chapter in global climate governance.

6.2.2. Cooperation with developed countries

Developed countries, by virtue of their strong economic strength and advanced technology, still play a leading role in the formulation of international climate governance rules in the short term. China shall make concerted efforts with developed countries to build a global multilateral climate governance mechanism together. China held the China-US Climate-Smart/Low-carbon Cities Summit to discuss policy research and capacity building of low-carbon cities and innovative application of low-carbon technologies; developed cooperation in China-EU carbon trading capacity building projects; and signed memorandums of understanding on climate change with Germany, Australia, New Zealand, Sweden and Switzerland to promote bilateral cooperation in technology, research, energy conservation and renewable energy.[68] In 2017, Canada, China and the EU jointly held the 1st Ministerial Conference on Climate Action, injecting new political impetus into the multilateral process on climate change.[69]

6.2.3. Participation in other multilateral processes

China has carried out extensive practical cooperation with international organizations. Specifically, it signed a memorandum of understanding with the UN Environment Programme on strengthening South-South cooperation on climate change; actively participated in relevant work of the IPCC; cooperated with the World Bank on the "Partnership for Market Readiness" program, jointly launched the Global Environment Facility and adopted the "Promotion of Clean and Green Cities in China Through International Cooperation" program; signed a bilateral memorandum of understanding on climate change cooperation with the Asian Development Bank; and established the alliance relationship with the International Energy Agency to strengthen cooperation in energy security, energy data and statistics, and energy policy analysis. Meanwhile, China also participated in a number of national scientific research programs under the framework of the Earth System Science Partnership to jointly advance basic research on global climate change and attended the Conference of the Parties to the Montreal Protocol on Substances that Deplete the Ozone Layer and the marine greenhouse gas emission reduction negotiations of the International Maritime Organization to implement emission reduction actions in the non-fossil energy sector. In addition, China has participated in discussions on climate change-related topics under multilateral mechanisms, such as the Group 20 (G20), Asia-Pacific Economic Cooperation (APEC), BRICS countries and BASIC countries. For example, at the G20 Hangzhou Summit, China, as the host called on all parties to jointly implement the *Paris Agreement* and discuss such major topics as climate change, clean energy and green finance. With the APEC High-level Roundtable on Green Development in Tianjin as a platform, China launched cooperation initiatives on global green supply chains and value chains and decided to jointly address global climate change challenges. As one of BRICS countries, China held a ministerial meeting on climate change every year with Brazil, South Africa and India, issued joint statements and established expert exchange mechanisms.

7. CONCLUSION

7.1. China's experience in addressing climate change

China has integrated climate change response with social and economic development goals and established a systematic climate policy system including planning and design, action plans, supervision, assessment and evaluation. First, climate goals were set to carry out reforms and define peak targets, and unit energy consumption and unit carbon emission binding indicators were included in national economic development plans to achieve two-way control of total consumption and intensity. Second, the Ministry of Ecology and Environment was established to form a governance system with overall planning and interdepartmental coordination and build a systematic climate policy system around mitigation and resilience issues, providing institutional guarantee for addressing climate change. Third, multilevel pilot areas were established to explore a new bottom-up climate governance pattern and covered cities, industrial parks and communities, becoming the main front for addressing climate change.

China has promoted the win-win cooperation in global climate governance and led the international community in jointly building multilateral climate governance mechanisms. China has always adhered to "common but differentiated responsibilities" based on national and global conditions, advocated to build a community with a shared future for mankind, adapted to global climate governance situations, promptly changed its roles, undertaken international obligations, and strived for development rights and interests. It has taken the lead in signing climate change agreements and implementing emission reduction targets, and set a good example for global response to climate change with its remarkable achievements in energy conservation, emission reduction and green development. China has developed all-round, multilevel cooperation on climate governance with developed countries, developing countries and international organizations, promoted political consensus on national climate change as well as exchange and application of science and technology, and shared benefits of ecological conservation with people around the world.

7.2. Challenges for China's response to climate change

In terms of global climate governance, China still faces arduous development tasks and needs to further accumulate strength to lead global climate governance. China is still a developing country in the primary stage of socialism, with insufficient conditions in all aspects; for example, there are still many problems to be solved, ecological conservaiton construction needs to be strengthened and consolidated, and its overall national strength remains to be enhanced. Therefore, it doesn't have the strength to lead global climate governance alone. In the reconstruction of top-level design for of future global climate governance, China, as the largest carbon emitter, will face more international responsibilities and obligations.

As for China's actions against climate change, continuous advancement of industrialization and urbanization increased the demand for carbon emissions, which increased the pressure of emission reduction and reduced the space for development. Due to uncompleted industrialization process and accelerated urbanization process in China, there is still room for further improvement in urbanization level and social consumption demand. Energy consumption and carbon emission demand caused by improved quality of residents' life will also continue to increase, adding pressure to urban emission reduction.

With respect to policy means and institutional guarantee for addressing climate change, we shall give full play to the role of market mechanism and further improve institutional construction. At the national level, there is a lack of legislation to address climate change, and the carbon market construction and the low-carbon development strategy are still not on the legal track; at the local level, there is still no detailed peaking roadmap by industry and sector, and the path of emission reduction remains to be clear. The market means of emission reduction are still simple. Specifically, carbon emission permit trading market, resource tax and other market means have just started, without any incentive mechanism for emission reduction, which makes it difficult to realize fundamental transformation of production mode.

7.2.1. China's outlook on addressing climate change

Through a series of reforms, China's economic development pattern has become greener and more low-carbon. New energy and new technologies will become new economic growth areas, with a huge potential in emission reduction. The Report to the Nineteenth CPC National Congress emphasized to "be good friends to the environment, cooperate to tackle climate change." In the new era, China shall continue to address climate change. For one thing, efforts shall be made to deepen the concept of green development, improve China's independent innovation capacity, accelerate the pace of economic transformation, and completely break the path dependence of economic growth on resource and energy consumption. For another, China shall shoulder more international responsibilities within its capacity, strive to build a multilateral mechanism for global climate governance and take the lead in global response to climate change.

From now to 2020, China formulated a peaking roadmap by sector, industry and field, with the focus on controlling total carbon dioxide emissions and intensity, so as to build a green, low-carbon and circular economic system gradually.

From 2020 to 2035, China will include climate change response, low-carbon development, and environmental pollution prevention and control into the unified framework of ecological advancement, with the aim to promote advanced experience nationwide and form an emission reduction pattern with Chinese characteristics. It will implement nationally determined contributions in an all-round way, so as to deepen international climate cooperation and consolidate its influence guided by the principles of "inclusiveness, cooperation, mutual trust and win-win progress."

Between 2035 and 2050, China will form a modern green economic system, occupy the top of global industrial chains and achieve absolute reduction in carbon emission after peaking. It will play a leading and coordinating role in global governance, share governance experience with the rest of the world and promote global eco-environmental advancement.

Notes

1. Third National Assessment Report on Climate Change.
2. Zhang Haibin (2007). China and International Climate Change Negotiations, *The Journal of International Studies*, No. 1.
3. Ding Yihui and Wang Huijun (2016). A New Understanding of Climate Change Science in China during the Last Century, *Chinese Science Bulletin*, No. 10.
4. Li Junfeng, Chai Qimin, Ma Cuimei, Wang Jijie, Zhou Zeyu and Wang Tian (2016). China's Climate Change Response Policy and Market Outlook, *Energy of China*, No. 1.
5. IEA, CO_2 *Emissions from Fuel Combustion* (online data service 2017 edition), http://www.iea.org/statistics/relateddatabases/co2emissionsfromfuelcombustion/.
6. *China's Policies and Actions for Addressing Climate Change* (2008 Annual Report).
7. Xue Jinjun and Zhao Zhongxiu, eds. (2012). Blue Book of Low-Carbon Economy: China's Low-Carbon Economy Development Report (2012). China: Social Sciences Academic Press.
8. Calculated based on relevant data from BP Statistical Review of World Energy (2017).
9. Calculated based on relevant data from Statistical Communiqué of the People's Republic of China on the 2017 National Economic and Social Development and BP Statistical Review of World Energy (2018).
10. First Biennial Update Report on Climate Change of China.
11. Du Yueying (2017). Bonn Relay of Climate Negotiations, *China Development Observation*, No. 22.
12. First Biennial Update Report on Climate Change of China.
13. China's Policies and Actions for Addressing Climate Change (2017 Annual Report).
14. IPCC Fifth Assessment Report: *Climate Change 2014 Synthesis Report*, http://www.ipcc.ch/report/ar5/syr/.
15. Qin Dahe. Proposal for Legislation to Address Climate Change, People's Daily Online, http://news.sina.com.cn/c/2009-03-11/192017386987.shtml, Accessed: March 11, 2009.
16. First Biennial Update Report on Climate Change of China.
17. Xie Zhenhua. China Will Invest in 30 Trillion Yuan to Address Climate Change in the Next 15 Years, Chinanews.com, http://finance.ifeng.com/a/20160423/14341545_0.shtml, accessed: April 23, 2016.
18. Zhang Xuefei and Xing Jianqiao (2015). South-South Cooperation Plays a Significant Role in Addressing Climate Change, *China Mining News*, December 9, 2015, p. 2.
19. First Biennial Update Report on Climate Change of China.
20. China's Scientific & Technological Actions on Climate Change.
21. China's Scientific and technological Actions on Climate Change.
22. China's Policies and Actions for Addressing Climate Change (2016 Annual Report).
23. China's Policies and Actions for Addressing Climate Change (2017 Annual Report).
24. China's National Climate Change Program.
25. Guiding Opinions on Resolving Serious Production Overcapacity Conflicts.

26 Xie Zhenhua (2014). Develop Circular Economy to Promote Green Transformation, *Economic Daily*, December 3, 2014, p. 3.
27 China Statistical Yearbook (2013–2016).
28 Report on the Work of the Government 2018.
29 China's Climate Change Response Policies and Actions (2017 Annual Report).
30 First Biennial Update Report on Climate Change of China.
31 First Biennial Update Report on Climate Change of China.
32 China Statistical Yearbook (2005–2011).
33 Report on the Work of the Government 2018.
34 Opinions on Deepening Development of Green Transportation.
35 China's Policies and Actions for Addressing Climate Change (2008 Annual Report).
36 Key Points for Climate Change Response Actions in Forestry during the Thirteenth Five-Year Plan Period.
37 First Biennial Update Report on Climate Change of China.
38 The latest report of UNFAO: China Ranks First in Net Increase of Forest Area in the Past Five Years in the World", November 5, 2015, gov.cn, http://www.gov.cn/xinwen/2015-11/04/content_2960277.htm.
39 Sponge pilot cities in 2015: Qian'an, Baicheng, Zhenjiang, Jiaxing, Chizhou, Xiamen, Pingxiang, Ji'nan, Hebi, Wuhan, Changde, Nanning, Chongqing, Suining, Gui'an New Area and Xi'xian New Area; sponge pilot cities in 2016: Fuzhou, Zhuhai, Ningbo, Yuxi, Dalian, Shenzhen, Shanghai, Qingyang, Xining, Sanya, Qingdao, Guyuan, Tianjin, Beijing.
40 The second batch of "urban repair and ecological restoration" pilot cities on April 18, 2017: Fuzhou, Xiamen and Quanzhou, Fujian Province; Zhangjiakou, Hebei Province; Kaifeng and Luoyang, Henan Province; Xi'an and Yan'an, Shaanxi Province; Nanjing, Jiangsu Province; Ningbo, Zhejiang Province; Harbin, Heilongjiang Province; Jingdezhen, Jiangxi Province; Jingmen, Hubei Province; Hulun Buir and Ulanhot, Inner Mongolia Autonomous Region; Guilin, Guangxi Zhuang Autonomous Region; Anshun, Guizhou Province; Xining, Qinghai Province; Yinchuan, Ningxia Hui Autonomous Region. The third batch of "urban repair and ecological restoration" pilot cities on July 14, 2017: Baoding and Qinhuangdao, Hebei Province; Baotou, Arxan, Hinggan League, Inner Mongolia Autonomous Region; Anshan, Liaoning Province; Fuyuan, Heilongjiang Province; Xuzhou, Suzhou, Nantong, Yangzhou and Zhenjiang, Jiangsu Province; Huaibei and Huangshan, Anhui Province; Sanming, Fujian Province; Ji'nan, Zibo, Jining and Weihai, Shandong Province; Zhengzhou, Jiaozuo, Luohe and Changyuan County, Henan Province; Qianjiang, Hubei Province; Changsha, Xiangtan and Changde, Hunan Province; Huizhou, Guangdong Province; Liuzhou, Guangxi Zhuang Autonomous Region; Haikou, Hainan Province; Zhunyi, Guizhou Province; Kunming, Baoshan, Yuxi and Dali, Yunnan Province; Baoji, Shaanxi Province; Golmud, Qinghai Province; Zhongwei, Ningxia Hui Autonomous Region; Urumqi, Xinjiang Uygur Autonomous Region.
41 Hohhot, Inner Mongolia Autonomous Region;Dalian, Liaoning Province; Chaoyang; Liaoning Province; Lishui, Zhejiang Province; Hefei, Anhui Province; Huaibei, Anhui Province; Jiujiang, Jiangxi Province;Jinan, Shandong Province; Anyang, Henan Province; Wuhan, Hubei Province; Shiyan, Hubei Province; Changde, Hunan Province; Yueyang, Hunan Province; Baise, Guangxi Zhuang Autonomous Region; Haikou, Hainan Province; Bishan District,Chongqing; Tongnan District, Chongqing; Guangyuan, Sichuan Province; Liupanshui, Guizhou Province and Bijie, Guizhou Province (Hezhang County); Shangluo, Shaanxi

Province; Xixian New District of Shaanxi Province; Baiyin, Gansu Province; Qingyang(Xifeng District), Gansu Province; Xining (Huangzhong County),Qinghai Province; Korla, Xinjiang Uygur Autonomous Region; Aksu(Baicheng County), Xinjiang Uygur Autonomous Region; Shihezi, Xinjiang Production and Construction Corps.
42 Zhang Jing (2015). Agricultural Measures for Addressing Climate change Must Be More And More Effective, *Science and Technology Daily*, December 27, 2015, p. 2.
43 China's Policies and Actions for Addressing Climate Change (2008 Annual Report).
44 Huang Junyi (2017). Chinese Miracle in Desertification Control, *Economic Daily*, September 11, 2017, p. 1.
45 According to the Outline of Implementation Plan for Pilot Areas of National Park System of the National Development and Reform Commission in 2015, pilot areas were set in Beijing, Jilin, Heilongjiang, Zhejiang, Fujian, Hubei, Hunan, Yunnan and Qinghai, and program lasted for three years.
46 China's Policies and Actions for Addressing Climate Change (2009 Annual Report).
47 China's Policies and Actions for Addressing Climate Change (2017 Annual Report).
48 China's Policies and Actions for Addressing Climate Change (2015 Annual Report).
49 China's Policies and Actions for Addressing Climate Change (2016 Annual Report).
50 National Strategy for Climate Change Resilience.
51 China's Policies and Actions for Addressing Climate Change (2016 Annual Report).
52 The Circulation on Establishing Low-Carbon Pilot Provinces and Cities, website of the National Development and Reform Commission, July 19, 2010, http://qhs.ndrc.gov.cn/dtjj/201008/t20100810_365271.html. The pilot areas included five provinces, namely Guangdong, Liaoning, Hubei, Shaanxi and Yunnan, and eight cities, namely Tianjin, Chongqing, Shenzhen, Xiamen, Hangzhou, Nanchang, Guiyang and Baoding.
53 The Circulation on Establishing the Second Batch of Low-Carbon Pilot Provinces and Cities, website of the National Development and Reform Commission, December 6, 2012, http://qhs.ndrc.gov.cn/dtjj/201008/t20100810_365271.html. The pilot areas included Beijing, Shanghai, Hainan Province, Shijiazhuang, Qinhuangdao, Jincheng, Hulun Buir, Jilin, Daxing'anling Region, Suzhou, Huaian, Zhenjiang, Ningbo, Wenzhou, Chizhou, Nanping, Jingdezhen, Ganzhou, Qingdao, Jiyuan, Wuhan, Guangzhou, Guilin, Guangyuan, Zhunyi, Kunming, Yan'an, Jinchang, Urumqi.
54 The Circulation on Establishing the Third Batch of Low-Carbon Pilot Cities, website of the National Development and Reform Commission, January 24, 2017, http://www.gov.cn/xinwen/2017-01/24/content_5162933.htm. The pilot aeras included forty-five cities (districts), namely Wuhai, Shenyang, Dalian, Chaoyang, Xunke County, Nanjing, Changzhou, Jiaxing, Jinhua, Quzhou, Hefei, Huaibei, Huangshan, Lu'an, Xuancheng, Sanming, Gongqingcheng, Ji'an, Fuzhou, Ji'nan, Yantai, Weifang, Changyang Tujia Autonomous County, Changsha, Zhuzhou, Xiangtan, Chenzhou, Zhongshan, Liuzhou, Sanya, Qiongzhong Li and Miao Autonomous County, Chengdu, Yuxi, Simao District of Pu'er, Lhasa, Ankang, Lanzhou, Dunhuang, Xining, Yinchuan, Wuzhong, Changji, Yining, Hetian, Alar City, the First Division of Xinjiang Production and Construction Corps.
55 Trial Exanimation for Greenhouse Gas Emission Targets of the National Development and Reform Commission to Test, February 16, 2014, People's Daily Online, http://legal.people.com.cn/n/2014/0216/c188502-24372967.html.
56 Wang Ke, Chen Mo (2018). Review and Prospect of Carbon Trading Market in China, *Transactions of Beijing Institute of Technology (Social Science Edition)*, No. 2.

57 Gong Xin (2017). From Stable Progress to Tightening, *China Economic Herald*, December 22, 2017, page B5.
58 Yi Lan, Li Chaopeng and Yang Li, et al. (2018). A Comparative Study on Development Degree of Seven Carbon Trading Pilot Areas in China, *China Population—Resources and Environment*, No. 2.
59 Zhu Yan (2014). Research on China's Climate Diplomacy, Doctoral dissertation of the Party School of the CPC Central Committee.
60 Xiao Lanlan (2016). China's Participation, Impact and Subsequent Actions for IPCC Assessment Report, *International Outlook*, No. 2.
61 The Umbrella Group was formed in 1997 and composed of the US, Canada, Japan, Australia, New Zealand, Norway, Iceland, Russia and Ukraine.
62 Su Wei, Lv Xuedu and Sun Guoshun (2008). Core Content and Prospect of Future UN Climate Change Negotiations: Interpretation of the Bali Roadmap, *Advances in Climate Change Research*, No. 1.
63 Zhong Sheng: China's Contributions to Global Climate Governance, *People's Daily*, November 26, 2015, p. 2.
64 Xi Jinping (2015). Work Together to Build a Win-Win, Equitable and Balanced Governance Mechanism on Climate Change, December 1, 2015, People's Daily Online, http://politics.people.com.cn/n/2015/1201/c1024-27873625.html.
65 Xi Jinping (2017). Work Together to Shoulder Responsibilities of The Times and Promote Global Development, February 18, 2017, People's Daily Online, http://politics.people.com.cn/GB/n1/2017/0118/c1001-29030932.html.
66 Xi Jinping (2015). Work Together to Build a Win-Win, Equitable and Balanced Governance Mechanism on Climate Change, December 1, 2015, People's Daily Online, http://politics.people.com.cn/n/2015/1201/c1024-27873625.html.
67 Positive Progress in South-South Cooperation on Addressing Climate Change, January 28, 2016, website of the National Development and Reform Commission, http://qhs.ndrc.gov.cn/qhbhnnhz/201601/t20160128_773390.html.
68 First Biennial Update Report on Climate Change of China.
69 China's Policies and Actions for Addressing Climate Change (2017 Annual Report).

CHAPTER SEVEN

Sustainable Development: Strategy and Practice

CHEN YING[*]

1. INTRODUCTION

"Sustainable development is development that meets the needs of the present without compromising the ability of future generations to meet their own needs." As a new approach to development, sustainable development is produced in response to the changes of the times and the needs of social and economic development, and resulted from people's reflection on the road that the mankind has gone through since the period of industrial civilization. The modern concept of sustainable development originated in the West and could be traced back to the warnings to the mankind, including the *Silent Spring*, the theory of "Spaceship Economy" and the Club of Rome in the 1960s. In 1972, the United Nations held the Conference on the Human Environment in Stockholm to meet the emerging

---

[*] Chen Ying, Doctor of Engineering of Tsinghua University, Research Fellow of Research Office on Sustainable Development Economics of Institute for Urban and Environmental Studies (now renamed as Research Institute for Urban Eco-civilization) of Chinese Academy of Social Sciences, Deputy Director of Research Center for Sustainable Development of Chinese Academy of Social Sciences, Professor and Doctoral Supervisor of Graduate School of Chinese Academy of Social Sciences, main author of the third working group of the 5th and 6th assessment reports of IPCC, his research fields include environmental economics and sustainable development, international climate governance, energy and climate policy.

global trend of environmental protection, which was an important symbol of the rise of global environmental protection and sustainable development movement. Since the Conference on the Human Environment in Stockholm in 1972, global sustainable development has gone through a bumpy course for more than forty years. The course is highlighted by many important milestones, such as *Our Common Future* published in 1987, UN Conference on Environment and Development in Rio in 1992, World Summit on Sustainable Development (Rio+10) in Johannesburg in 2002, United Nations Conference on Sustainable Development (Rio+20) in 2012, and *2030 Agenda for Sustainable Development* adopted at the UN General Assembly in New York in 2015. Sustainable development has developed from a concept, gradually gained popular support and become a development strategy and practice of all countries.

The modern idea of sustainable development originated from the West and was later imported to China. However, during the reflection of its own development, China gradually accepted the concept of sustainable development and resonated with it, and then promoted sustainable development as a national strategy and actively implemented the sustainable development strategy. In the past forty years of reform and opening-up, China has not only witnessed and experienced the global process of sustainable development but also actively promoted the sustainable development strategy as a major developing country and made great contributions to the global process of sustainable development. The review of this process and the summary of the regularity and features of China's development and evolution provide a reflection of China's development and changes over the past forty years of reform and opening-up. Meanwhile, it also inspires people to think deeply about how China could contribute its wisdom to global sustainable development in the new era, promote win-win cooperation with the thought on ecological civilization and build a community with a shared future for mankind in the future.

2. ENVIRONMENTAL PROTECTION: ORIGIN OF SUSTAINABLE DEVELOPMENT (1972–1991)

In 1972, China was in the midst of the "Great Cultural Revolution," with a low economic development level in the early stage of industrialization. Although environmental pollution appeared in some places, most people still knew little about environmental problems and even thought that environmental problems were specific to capitalist countries and there was no pollution in socialist countries. China, which was a relatively closed country, began to recognize environmental issues after participating in an international conference, which started a unique process of environmental protection in China. Though the environmental protection work was still in its infancy at that time, without in-depth understanding and a complete system, this enlightenment stage of environmental protection laid an important foundation for spreading the idea of sustainable development ideas and establishing the sustainable development strategy subsequently.

2.1. Environmental protection has been put on national political agenda

In 1972, the Chinese government decided to assign a delegation to the UN Conference on the Human Environment held in Stockholm. This was the first major international conference that China attended since the restoration of China's seat in the UN, and opened an important window for China to understand its own environmental problems. The delegates attending the conference came to recognize the severity of environmental problems, and realized that Chinese cities had even severer environmental pollutions than western countries and a much serious natural ecological damage beyond western countries, instead of "no pollution in socialist countries." In August 1973, China held the first conference on environmental protection, which was attended by more than 300 representatives, including heads of local governments and relevant ministries and commissions, factory representatives and scientists. At the late session of the conference, Premier Zhou decided to hold a 10,000-person congress at the Great Hall of the People to popularize the environmental protection awareness in the whole society. The conference played a very important role in raising the environmental awareness in all walks of life, and adopted China's "32-character guidelines" on environmental protection, involving "comprehensive planning, rational layout, comprehensive utilization, waste recycling, mobilization of the masses, universal participation, environmental protection and people's well-being." It also adopted the *Regulations on Environmental Protection and Improvement*, put forward requirements for environmental protection, and made arrangements.

2.2. Environmental protection has become a basic state policy

After the 1st National Conference on Environmental Protection, the Leading Group of Environmental Protection under the State Council (with a general office under it) urged local governments to set up relevant environmental protection agencies to investigate and assess environmental pollutions and conduct environmental governance focusing on smoke prevention and dust control. In 1983, the 2nd National Conference on Environmental Protection was held. Li Peng, then Vice Premier, announced on behalf of the State Council to establish environmental protection as a basic state policy of China, so as to move environmental protection from the edge of economic construction to the center. In order to implement this basic state policy, the State Council formulated and issued the "synchronous development" guidelines, abandoned the old road of "pollution first, environmental governance later" and required "simultaneous planning, implementation and development of economic construction, urban and rural construction and environmental construction to achieve unification of economic, social and environmental benefits." Based on China's specific problems and demands, these strategic guidelines were in harmony with coordinated economic, social and environmental development pursued by the western idea of sustainable development.

2.3. A legal system and policy system for environmental protection has been initially established

In order to implement the basic state policy of environmental protection, China promulgated the *Environmental Protection Law* in 1979, marking that China's environmental protection moved onto the legal track. In 1989, China revised the *Environmental Protection Law*. Meanwhile, it successively formulated and promulgated the *Law on the Prevention and Control of Water Pollution*, the *Law on the Prevention and Control of Air Pollution*, the *Law on the Prevention and Control of Air Pollution* and other specific laws and standards on pollution prevention and control, and issued the *Forest Law*, the *Grassland Law*, the *Water Law*, and the *Water and Soil Conservation Law*, the *Wildlife Protection Law* and other resources protection laws, initially constituting a legal framework of environmental protection. In 1988, the Environmental Protection Bureau was separated from the Ministry of Urban and Rural Construction and Environmental Protection and became independent Environmental Protection Administration directly under the State Council, marking that environmental protection was formally included in the government's administrative system. In 1989, the 3rd National Conference on Environmental Protection was held and put forward three major policies for environmental protection, namely prevention-based pollution control, polluter-oriented pollution control and intensified environmental management. Meanwhile, eight management systems were issued, namely the "three-simultaneity" system, environmental impact assessment system, pollution charging system, comprehensive urban environmental renovation and quantitative assessment system, target responsibility system of environmental protection, pollution discharge reporting and registration and pollution discharge permit system, deadline governance system and centralized pollution control system, preliminarily building the environmental protection policy system.

3. THE ESTABLISHMENT AND IMPLEMENTATION OF NATIONAL SUSTAINABLE DEVELOPMENT STRATEGY (1992–2000)

In 1992, China's reform and opening-up, and economic construction were in full swing. In particular, disorderly development of township enterprises extended industrial pollution from cities to villages, making environmental problems increasingly prominent. Another important international conference exerted a profound influence on China's development course, which made the concept of sustainable development deeply rooted in the people's minds, helped establish China's national sustainable development strategy and turned sustainable development from an idea into policies and actions.

3.1. Sustainable development evolved from a concept into a national strategy

From June 3 to 14, 1992, the UN Conference on Environment and Development (hereinafter referred to as the "Rio Conference") was held in Rio de Janeiro, Brazil. Against the backdrop of continuous global environmental degradation and increasingly serious development problems, the Rio Conference was held as another high-level international meeting subsequent to the UN Conference on the Human Environment held in Stockholm, Sweden, in June 1972. After more than two years of preparations and consultations, the conference adopted the *Rio de Janeiro Declaration on Environment and Development* (also known as the *Earth Charter*), the *Agenda 21* and the *Statement of Principles on Forests*; and signed the UN Framework Convention on Climate Change and the UN Convention on Biological Diversity. The Chinese government attached great importance to the conference. Under the leadership of the Ministry of Foreign Affairs, the State Science and Technology Commission, the State Planning Commission and the State Environmental Protection Administration constituted a preparatory group and actively participated in various preparatory activities organized by the UN. In the preparation process of the Rio Conference, developing countries formed the consultation mechanism of the "G77 and China," which was later widely used in many international negotiations. The Chinese government sent a high-level delegation to attend the Rio Conference. Li Peng, then Prime Minister of the State Council, attended the summit conference and delivered an important speech, explicitly expounded China's five proposals for strengthening international cooperation on environment and development, took the lead in signing two important conventions on climate change and biodiversity on behalf of the Chinese government among the five powers, and made a solemn commitment for fulfilling the

Agenda 21 and other documents, fully demonstrating China's great importance and sense of responsibility to global environment and development issues.

Shortly after the conference, the CPC Central Committee and the State Council issued the Ten Countermeasures for Environment and Development,[1] which started with a proposal for "implementing the sustainable development strategy." According to the document, "China's economic development still followed the traditional development pattern characterized by massive consumption of resources and extensive management, which not only caused a great damage to the environment but also was hard to sustain the development itself. Therefore, changing the development strategy and turning to the road of sustained development were the right choice to accelerate China's economic development and solve environmental problems." Obviously, "sustained development" mentioned above actually referred to "sustainable development," which was the first time that China put forward the implementation of sustainable development strategy.

From then on, sustainable development has been strengthened as a national strategy. For example, in his speech at the 5th Plenary Session of the 14th CPC Central Committee in September 1995, then President Jiang Zemin pointed out that "sustainable development must be taken as a major strategy in the process of modernization." In March 1996, the fourth session of the Eighth National People's Congress (NPC) approved the *Report of the People's Republic of China on the Outline of the Ninth Five-Year Plan for National Economic and Social Development and the Long-range Objectives through the Year 2010*[2] which included the sustainable development strategy in the national plan for the first time. In March 2001, the fourth session of the Nineth NPC adopted the *Outline of the Tenth Five-Year Plan*, giving top priority to the implementation of the sustainable development strategy and completing the historic process from the establishment of the sustainable development strategy to the comprehensive promotion. On July 1, 2001, Chinese President Jiang Zemin explained China's sustainable development strategy in an all-round way at the Party's 80th anniversary meeting, stressing to "adhere to the sustainable development, properly balance the relationship between economic development as well as population, resources and environment, improve the ecological environment and beautify the living environment, and improve public facilities and social welfare facilities, in an attempt to blaze a path of civilized development featuring increased production, higher living standards, and healthy ecosystems."

3.2. China took the lead in formulating *China's Agenda 21*

After the Rio Conference, in July 1992, the Environmental Protection Commission of the State Council organized more than 300 experts from fifty-two ministries, commissions and institutions to compile *China's Agenda 21* based on *Agenda*

21[3] as the overall strategy, plan and measures for China's sustainable development. After more than a year's efforts, on March 25, 1994, *China's Agenda 21* was discussed and adopted at the 16th Executive Meeting of the State Council,[4] which was also known as *White Paper on China's Population, Environment and Development in the 21st Century*. It was a guidance document for the Chinese government to formulate medium- and long-term plans for national economic and social development, as well as the world's first national-level sustainable development strategy.

China's Agenda 21 is divided into four parts, and each part contains many chapters, with a total of twenty chapters. Each chapter includes an introduction and a series of program areas, totaling seventy-eight program areas. Specifically, the first part is the overall strategy and policy of sustainable development, discussing the background and necessity of proposing China's sustainable development strategy and putting forward short-term, medium-term and long-term goals of China's sustainable development strategy.

(1) The short-term goal (1994–2000) focused on taking emergency actions for solving China's existing prominent contradictions between environment and development, and laying a solid foundation for developing major measures for long-term sustainable development, so as to prevent environmental quality, quality of life and status of resources from further degrading and make local improvements, while maintaining an economic growth rate of around 8% in China; another key short-term goal was to strengthen sustainable development capacity building.

(2) The medium-term goal (2000–2010) focused on taking a series of sustainable development actions for changing development pattern and consumption pattern, and improving the management system, economic and industrial policies, technological system and social norms that adapted to sustainable development.

(3) The long-term goal (since 2010) focuses on restoring and improving the regulatory function of China's economic ecosystem, keeping China's economic and social development within the carrying capacity of its environment and resources, exploring a modern road of efficient, harmonious and sustainable development suitable for China's national conditions, and making due contributions to the global sustainable development process.

China's Agenda 21 also proposes specific action objectives and policy measures in terms of social and economic sustainable development, and rational usage of resources and environmental protection. Sustainable social development includes population, consumer spending and social services, poverty eradication, health and wellness, human settlements, and disaster prevention and mitigation. Sustainable economic development takes the promotion of rapid economic growth as a necessary condition for eliminating poverty, improving people's living standards and enhancing comprehensive national strength, including economic policy for sustainable development, sustainable development of agriculture and rural economy, sustainable development of industry, transportation and communication industry, sustainable energy and production and consumption. There is also rational usage of resources and environmental protection, including protection and sustainable usage of water, soil and other natural resources, biodiversity protection, land desertification prevention and control, disaster prevention and mitigation, atmosphere protection, air pollution control and acid rain prevention, and biosafety proposal of solid waste.

3.3. The sustainable development strategy was fully implemented

Since the release of *China's Agenda 21*, its implementation has set off an upsurge of sustainable development in China. First, a large number of studies on sustainable development emerged in the academic world. For example, on January 4, 1994, China Sustainable Development Research Center of Peking University was established, and Mr. Ye Wenhu delivered a keynote report entitled *Creating a New Civilization of Sustainable Development* at the conference and published a collection of essays on the *Road to Sustainable Development*.[5] Second, relevant organizational mechanisms were established. On March 25, 1994, the Ministry of Science and Technology established the Administrative Center for China's Agenda 21 to promote the management and implementation of *China's Agenda 21* and its priority projects. In August 2000, the State Council approved the establishment of the national leading group for promoting sustainable development. The leading group was headed by the State Planning Commission, with the Ministry of Science and Technology as the deputy headed and eighteen departments of the State Council as participating units. Third, all local governments and departments took action to incorporate the requirements of *China's Agenda 21* into local and sectoral development plans,[6] and launched many agenda 21 agendas at local and sectoral levels. At the local level, *Sichuan's Agenda 21* was compiled in 1995; due to the adjustment of Sichuan's administrative divisions, local government of some prefecture-level cities in Sichuan Province, such as Chengdu and Panzhihua, also compiled *Chengdu Agenda 21* and *Panzhihua Agenda 21* and their priority projects. At the industry level, *China's Ocean Agenda 21* was compiled by the State Oceanic Administration in 1996, *China's Paper Industry Agenda 21* was compiled in 1997 and *China's Agenda 21 on Agricultural Action Plan* was compiled in 1999. Fourth, pilot programs were carried out to draw upon the experience and promote the work in all areas. In November 1997, in order to strengthen the guidance for local implementation of *China's Agenda 21* and further promote the implementation of local agenda 21, the State Planning Commission and the State Science and Technology Commission issued a document jointly to carry out the local pilot work of *China's Agenda 21* in Beijing, Hubei, Guizhou, Shanghai, Hebei, Shanxi, Jiangxi, Sichuan and other provinces and municipalities, as well as Dalian, Harbin, Guangzhou, Changzhou, Benxi, Nanyang, Tongchuan, Chizhou and other cities, with the aim to gradually accumulate experience to further promote China's and local *Agenda 21*. The comprehensive experimental area for social development, which was launched as early as in 1986, was renamed the "National Experimental Area for Sustainable Development" in December 1997; by the end of 2016, 189 experimental areas had been established, becoming bases for implementing *China's Agenda 21* and the sustainable development strategy. Fifth, the progress of sustainable development was summarized in a timely manner to fulfill international obligations. In June 1997, China submitted the *People's Republic of China National Report on Sustainable Development* to the Nineteenth Special Sessions of the UN General Assembly on Environment and Development, which consisted of five parts, summarizing the overall progress in sustainable development since 1992 and the specific actions, achievements, subsequent arrangements and measures in fourteen key areas, and expounding China's basic principles and positions on several international issues concerning sustainable development.

4. EFFORTS AND ACHIEVEMENTS IN IMPLEMENTING MILLENNIUM DEVELOPMENT GOALS (2001–2015)

4.1. Participation in formulating UN Millennium Development Goals

The 800-page *Agenda 21* contains more than 2,500 diverse action recommendations and is a non-legally binding action blueprint for a global sustainable development plan. However, putting these actions into practice remains a huge challenge. In September 2000, world leaders gathered in New York to attend the UN Millennium Summit; 189 countries signed the UN Millennium Declaration, promising to "spare no effort to free our fellow men, women and children from the abject and dehumanizing conditions of extreme poverty, to which more than a billion of them are currently subjected," and formulating eight goals to guide development in the next fifteen years, which were called the Millennium Development Goals (MDGs). Specifically, the eight goals are eradicating extreme poverty and hunger, achieving universal primary education, promoting gender equality and empower women, reducing child mortality, improving maternal health, combating HIV/AIDS, malaria and other diseases, ensuring environmental sustainability and developing a global partnership for development. Subsequently, the UN developed statistical indexes for measuring the progress of the MDGs and regularly published evaluation reports to promote the implementation of the MDGs. The 1st World Summit on Sustainable Development (WSSD) held in Johannesburg, South Africa, in 2002, comprehensively reviewed and evaluated the implementation of Agenda 21 and restated the efforts to achieve the MDGs. Prior to the conference, China submitted the *People's Republic of China National Report on Sustainable Development*, which summarized the progress in implementing sustainable development. Then Chinese President Jiang Zemin attended the conference and delivered a speech. He pointed out that we need a process of economic globalization featuring "win-win," equality, fairness and coexistence among all countries in the world, and the key to promote common development of mankind lies in a new international economic order that is just, fair and equitable. In the twenty-first century, China has made great efforts to achieve the MDGs and will follow the road of sustainable development more unswervingly.

4.2. Great achievements in implementing Millennium Development Goals

According to the *Millennium Development Goals Report 2015* released by the UN Department of Economic and Social Affairs in July 2015, "unprecedented efforts and far-reaching achievements" were made in implementing the MDGs.[7] According to the detailed data in the report, a significant progress was made on all the goals, including halving global extreme poverty population, accessing primary education based on gender equality and halving the number of people without access to improved drinking water. During the period, China not only made all-out efforts to implement the MDGs by integrating them into its own development plans but also scored remarkable achievements. Besides, as a responsible major developing country, China was active in providing support and assistance to other developing countries in achieving the MDGs. There was no doubt that China made great contributions to global sustainable development.

According to the *Report on China's Implementation of the Millennium Development Goals* (2000–2015) released in July 2015,[8] China has achieved or basically achieved thirteen indexes of the MDGs (as shown in Table 7-1 for details), especially in poverty reduction. Figure 7-1 shows the changes in human development index in China and the world from 1990 to 2013. In 2013, China exceeded the world's average and ranked among countries with a high human development index.

From 2000 to 2015, China had maintained a steady and fast economic growth. It recorded GDP growth from RMB10.0 trillion yuan in 2000 to RMB63.6 trillion yuan in 2014, ranking second in the world. The rapid economic development provided strong support to the increase of urban and rural incomes and the

Table 7-1 China's implementation of the MDGs

Specifical goals	Achievement
MDG 1: Eradicate extreme poverty and hunger	
MDG 1A: Halve, between 1990 and 2015, the proportion of people whose income is less than $1.25 a day	Already achieved
MDG 1B: Achieve full and productive employment and decent work for all, including women and young people	Basically achieved [1]
MDG 1C: Halve, between 1990 and 2015, the proportion of people who suffer from hunger	Already achieved
MDG 2: Achieve universal primary education	
MDG 2A: Ensure that, by 2015, children everywhere, boys and girls alike, will be able to complete a full course of primary schooling	Already achieved
MDG 3: Promote gender equality and empower women	
MDG 3A: Eliminate gender disparity in primary and secondary education, preferably by 2005, and in all levels of education no later than 2015	Already achieved
MDG 4: Reduce child mortality	
MDG 4A: Reduce by two-thirds, between 1990 and 2015, the under-five mortality rate	Already achieved
MDG 5: Improve maternal health	
MDG 5A: Reduce by three quarters, between 1990 and 2015, the maternal mortality ratio	Already achieved
MDG 5B: Achieve, by 2015, universal access to reproductive health	Basically achieved [2]
MDG 6: Combat HIV/AIDS, Malaria and other diseases	
MDG 6A: Have halted by 2015 and begun to reverse the spread of HIV/AIDS	Basically achieved [3]
MDG 6B: Achieve, by 2010, universal access to treatment for HIV/AIDS for all those who need it	Basically achieved [4]
MDG 6C: Have halted by 2015 and begun to reverse the incidence of malaria and other major diseases	Basically achieved [5]
MDG 7: Ensure environmental sustainability	
MDG 7A: Integrate the principles of sustainable development into country policies and programs and reverse the loss of environmental resources	Basically achieved [6]
MDG 7B: Reduce biodiversity loss, achieving, by 2010, a significant reduction in the rate of loss	Unachieved yet
MDG 7C: Halve, by 2015, the proportion of the population without sustainable access to safe drinking water and basic sanitation	Already achieved

(*continued*)

Table 7-1 Continued

Specifical goals	Achievement
MDG 7D: Achieve, by 2020, a significant improvement in the lives of at least 100 million slum dwellers	Possibly achieved
MDG 8: Develop a Global Partnership for Development	—

Note 1: By the end of 2014, China's total employment was 772.53 million people, and the registered urban unemployment rate was 4.09%. From 2003 to 2014, a total of 137 million new jobs were increased in urban areas.

Note 2: In 2013, free basic technical services for family planning covered 100% of registered population and 96% of migrant population, with the systematic management rate for pregnant women as high as 89.5%.

Note 3: In 2014, 104,000 new HIV infections and patients were reported, with an annual growth rate of 14.8%. The epidemic situation was generally kept at a low epidemic level, and the fast-rising HIV incidence was initially curbed. The fatality rate of AIDS patients meeting treatment criteria dropped from 33.1% in 2003 to 6.6% in 2013.

Note 4: China began to provide free voluntary HIV/AIDS counseling and testing services in 2004. By 2014, a HIV/AIDS prevention and treatment service network covering both urban and rural areas had been basically established.

Note 5: The rising trend of tuberculosis was effectively curbed in China, and the incidence of malaria was significantly reduced. However, the incidence of chronic diseases has been on the rise in recent years.

Note 6: Since 2000, China has fully incorporated the sustainable development principle into its national economic and social development plan. China's ecosystem has been improved in general, and the trend of continuous environmental deterioration has been initially curbed.

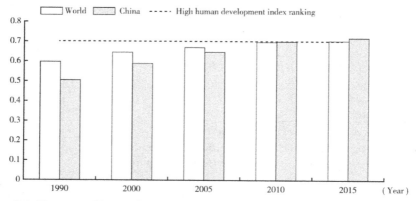

Figure 7-1 Changes in China and the world's average human development index from 1990 to 2013

Source: http://hdr.undp.org/en/content/table-2-human-development-index-trends-1980-2013.

acceleration of the poverty alleviation process. In 2014, the per capita disposable income of urban residents was RMB28,844, and the per capita net income of rural residents was RMB9,892, which were 3.59 times and 3.39 times of those in 2000, respectively. The number of poverty population in China dropped from 689 million in 1990 to 250 million in 2011, a decrease of 439 million, making a major contribution to global poverty reduction. Since 2004, China's grain output increased for eleven consecutive years, feeding nearly 20% of the world's population with less than 10% of the world's arable land. China made great efforts in promoting public health, education and other livelihood projects. Since 2000, a total of 467 million rural residents have gained access to safe drinking water, and the net enrollment rate of both male and female primary-school-age children has remained at over 99%. While realizing self-development, China has also actively carried out South-South cooperation and provided assistance to more than 120 developing countries to implement the MDGs to the best of its ability.

The report summarized China's main practices in implementing the MDGs, including: (1) give top priority to development, and keep innovating development concepts based on national conditions; (2) formulate and implement medium-term and long-term national development strategic plans, and fully incorporate the MDGs as binding targets into its national plans; (3) establish and improve legal and regulatory systems, and mobilize the extensive participation of all walks of life; (4) vigorously strengthen capacity building, and actively carry out experiments and demonstrations; (5) strengthen foreign development cooperation and promote mutual learning of development experience. Meanwhile, the report also pointed out that the gaps between China and developed countries should not be ignored, with an arduous poverty eradication task, incomplete development, and prominent imbalance and un-coordination between regions and between urban and rural areas.

5. A NEW JOURNEY TOWARD SUSTAINABLE DEVELOPMENT BY 2030 (2015–PRESENT)

The year 2015 was a crucial year for global sustainable development. As the MDGs expired in 2015, the direction of global sustainable development became a new issue facing politicians around the world. The UN General Assembly 2010 authorized to launch consultations and discussions on the Post-Millennium Development Goals (Post-MDGs) and the Post-2015 Development Agenda. The "Rio+20" UN Conference on Sustainable Development in 2012 decided to establish an open working Group to negotiate the sustainable development goals, while promoting consultations and discussions on Post-MDGs. After nearly three years of preparation, with the active promotion of the UN, the World Summit on Sustainable Development held in New York in September 2015 adopted *Transforming Our World: The 2030 Agenda for Sustainable Development*,[9] with a set of 169 specific targets in 17 fields at its core.[10] On January 1, 2016, the *2030 Agenda for Sustainable Development* officially entered the implementation stage. It is a programmatic document for guiding global sustainable development for the next fifteen years, marking a new chapter in global sustainable governance. As China implemented the *2030 Agenda for Sustainable Development* under the guidance of the thought on ecological civilization, China's sustainable development has entered a new era.

5.1. China's roles in negotiations on *Post-2015 Development Agenda*

In order to formulate the *Post-2015 Development Agenda*, various forms of discussions, consultations and negotiations were carried out at different levels in different scopes and fields, with complex interest disputes among various countries with different positions. China took an active part in relevant preparatory work and played positive and constructive roles. In September 2013, China's Ministry of Foreign Affairs published China's Position Paper on the Development Agenda beyond 2015,[11] which not only introduced China's practice in implementing the MDGs but also comprehensively expounded China's position and propositions on the *Post-2015 Development Agenda*. In particular, China proposed to focus on five basic principles, namely core topics, diversity in development model, "common but differentiated responsibilities," win-win cooperation and equal consultation. The three key fields proposed by China included (1) eradicating poverty and hunger, and promoting economic growth; (2) comprehensively promoting social progress and safeguarding equity and justice; (3) enhancing ecological conservation and promoting sustainable development. China stood for establishing a more equal and balanced global partnership for development, enhancing the development capacity of all countries and giving full play to the coordinating role of the UN. Meanwhile, all countries should incorporate the development agenda into their national development strategies based on their respective national conditions, with a certain policy space and flexibility, carry them out in the principle of voluntariness and evaluate the implementation. Top priority should be given to the supervision over implementing measures at the international level, with a focus on reviewing the implementation of official commitments to development assistance, technology transfer and capacity building. Emphasis should also be placed on helping developing countries strengthen data statistics capacity building and improve quality and timeliness of the data. China's position and propositions demonstrated to the international community China's attitude of actively, pragmatically and constructively participating in global sustainable development governance as a responsible major power.

5.2. China's vision for implementing *2030 Agenda for Sustainable Development*

The year 2015 marks the 70th anniversary of the Founding of the UN. On September 28, 2015, Chinese President Xi Jinping attended the UN Summit on Sustainable Development and delivered a speech entitled "Working Together to Build New Partnership of Win-Win Cooperation and to Build a Community with a Shared Future for Mankind,"[12] putting forward the development concepts of "fairness, openness, inclusiveness and innovation" and elaborated on China's views on global development issues. Furthermore, he called for renewing "our commitment to the purposes and principles of the UN Charter, build a new type of international relations featuring win-win cooperation, and create a community of shared future for mankind," and called on the international community to strengthen cooperation and jointly implement the *Post-2015 Development Agenda*, in order to strive for win-win cooperation. These development concepts were rooted in China's practice, reflected deep thinking and active interpretation of the development of all countries and the common problems facing the world, demonstrated Chinese wisdom in seeking a long-term development in the future together, and charted the right course for the common development of all countries. China also announced to establish the South-South Cooperation Assistance Fund, with the initial contribution of USD 2 billion to support developing countries in implementing the *Post-2015 Development Agenda*. Besides, China and the UN also co-hosted a roundtable on South-South cooperation during the summit, with the aim to promote the effectiveness of South-South cooperation under the framework of the *Post-2015 Development Agenda*.

Chinese leaders also stressed China's determination and confidence to work with the international community to implement the *2030 Agenda for Sustainable Development* on many international occasions. For example, Chinese President Xi Jinping delivered an important speech entitled "Innovative Growth That Benefits All" at the 10th summit of the Group of Twenty (G20) major economies in Antalya, Turkey, on November 15, 2015,[13] proposing "to implement the *2030 Agenda for Sustainable Development* and lend strong impetus to equitable and inclusive development." China is committed to lifting more than 70 million rural people out of poverty in the next five years, setting up the South-South cooperation assistance fund, continuing to increase investments in the least developed countries and supporting developing countries in implementing the *2030 Agenda for Sustainable Development*. China will incorporate the implementation of the *2030 Agenda for Sustainable Development* into its Thirteenth Five-Year Plan. In particular, Chinese President Xi Jinping also suggested that "all the G20 members come up with their own national plans of implementation which will be combined to form a G20 plan of action to boost strong, sustainable and balanced global growth."

In April 2016, the Ministry of Foreign Affairs took the lead in the world to publish *China's Position Paper on the Implementation of the 2030 Agenda for Sustainable Development*,[14] proposing six general principles of peaceful development, win-win cooperation, integration and coordination, inclusiveness and openness, sovereignty and voluntary action and "common but differentiated responsibilities," and defining the key areas and priorities of eradicating poverty and hunger, maintaining economic growth, advancing industrialization, improving social security and social services, safeguarding equity and justice, strengthening environmental protection, giving active response to climate change, making effective use of energy resources and improving national governance. On this basis, China put forward five suggestions for the ways of implementation to the international community. First, strengthen capacity building. The national government should take primary responsibility for its development. It is important to align national development strategies with the implementation of the *2030 Agenda for Sustainable Development*, making them mutually reinforcing. Second, create an enabling international environment for development. It is important for all countries to promote a balanced, win-win and inclusive multilateral trading system and form a fair, reasonable and transparent system of international economic and trading rules, and promote rational flow of production factors, efficient allocation of resources and deep integration of markets. Third, strengthen development partnership. The international community should work toward a more equitable and balanced global partnership for development and maintain North-South cooperation as the main channel for development cooperation. Fourth, promote coordination mechanism. Development policy should be incorporated into global macroeconomic policy coordination. G20 members are encouraged to formulate a meaningful and executable collective action plan for implementation of the *2030 Agenda*, so as to play a leading role in the implementation and complement the UN process. Fifth, improve follow-up and review. The UN High-Level Political Forum should play a key role in the follow-up and review.

5.3. Formulation of China's national plan on implementation of the 2030 agenda for sustainable development

The Chinese government attaches great importance to the implementation of the *2030 Agenda for Sustainable Development* by making work arrangements and formulating relevant policies rapidly. In October 2015, the 5th Plenary Session of the Eighteenth CPC Central Committee adopted the *Communique*, which put forward the five development concepts of innovation, coordination, green, openness and sharing during the Thirteenth Five-Year Plan period, and proposed to actively participate in the *2030 Agenda for Sustainable Development*.[15] Incorporating the *2030 Agenda for Sustainable Development* into the Thirteenth Five-Year Plan is China's basic strategy and thinking for implementing sustainable development goals. In October 2015, Chinese President Xi Jinping announced at the 2015 Global Poverty Reduction and Development Forum that in the next five years, China will lift more than 70 million people living below the current poverty line out of poverty. This is an important step in China's implementation of the *2030 Agenda for Sustainable Development*. On November 5, 2015, Zhang Jun, Director-General of the Department of International Economic Affairs of the Ministry of Foreign Affairs, chaired a coordination meeting to study the implementation work of the *2030 Agenda for Sustainable Development*. Heads of relevant departments of more than thirty units in China, including the National Development and Reform Commission, the Ministry of Finance and the Ministry of Commerce, attended the meeting. In-depth discussions were made at the meeting, focusing on strengthening the integration of the *2030 Agenda for Sustainable Development* and China's development plans, establishing implementation coordination mechanisms and implementation schemes in various fields, and carrying out international cooperation.[16]

On September 19, 2016, Chinese Premier Li Keqiang chaired and delivered an important speech at a symposium on the "Sustainable Development Goals: Working Together to Transform Our World—China's Proposal" at the United Nations Headquarters in New York.[17] He pointed out that sustainable development was based on development, without which nothing can be done. Development must be sustainable and coordinated among economy society and environment. Sustainable development is also open, interconnected and inclusive as a common cause of the world. The key to the implement the *2030 Agenda for Sustainable Development* is to take actions; considerations should be given to targets at present and in the long term, and priority shall be given to core tasks, in an attempt to make effective achievements as soon as possible. China put forward two priority areas: one was to eradicate poverty and hunger, and the other was to address unbalanced development. Both problems were urgent for developing countries to solve. In addition, refugees, public health, climate change and other prominent issues were mentioned as well. Chinese Premier Li Keqiang's speech demonstrated China's strategy and determination to implement the *2030 Agenda for Sustainable Development* to the international community once again. Meanwhile, he also briefed on China's progress in implementing the agenda and announced that the Chinese government had approved and would release *China's National Plan on Implementation of the 2030 Agenda for Sustainable Development*. China's National Plan focused on the interconnection of goals; all the specific goals set forth in the *2030 Agenda* will be included into the general national development plan, and refined, coordinated and connected in specialized plans. Efforts will be made to strengthen the implantation guarantees, establish and improve corresponding systems and mechanisms, mobilize the whole society to increase resource input, and strengthen supervision and evaluation. China is confident and capable of achieving all the goals on schedule. China is ready to take an active part in relevant international cooperation and support the UN in playing a bigger role in implementing the 2030 Agenda for Sustainable Development. In addition, China will continue to increase its input in South-South cooperation and share its development experience and opportunities with other countries. By the end of 2015, China had contributed over RMB400 billion to 166 countries and international and regional organizations and trained more than 12 million people in developing countries. By 2020, China's total annual contribution to relevant UN development agencies increased by USD 100 million from that in 2015.

On October 26, 2016, *China's National Plan on Implementation of the 2030 Agenda for Sustainable Development* was published on the official website of the Ministry of Foreign Affairs of China.[18] In addition to reviewing China's achievements and experience in implementing the MDGs, it analyzed the opportunities and challenges in implementing the 2030 Agenda for Sustainable Development, defined that China would carry out the work with the five development concepts of "innovation, coordination, green, openness and sharing" as the guiding ideology, adhere to the principles of peaceful development, win-win cooperation, integration and coordination, inclusiveness and openness, and focus on synergy of strategies, institutional guarantee, social mobilization, resource input, risk management, international cooperation, and oversight and review, and implement the *2030 Agenda for Sustainable Development* in an incremental way. The plan also decomposed the tasks in accordance with the sustainable development goals and elaborated on China's implementation of the 17 sustainable development goals and 169 specific targets in the future. It can be seen that China's strategy on implement on the *2030 Agenda for Sustainable Development* was divided into domestic actions and international cooperation, which was complementary with China's ongoing strategies of ecological conservation as well as green and low-carbon development. Instead of being implemented separately, the *2030 Agenda for Sustainable Development* should be internalized into China's Thirteenth Five-Year Plan, and extended and implemented by local governments and departments under the policy guidance. The *2030 Agenda for Sustainable Development* not only promoted global environmental governance but also provided a stage for China to play a bigger role in global environmental governance. In August 2017, China launched Center for International Knowledge on Development and released the *China's Progress Report on Implementation of the 2030 Agenda for Sustainable Development*,[19] which systematically reviewed China's progress in implementing the seventeen sustainable development goals since September 2015, as well as the challenges it faces and plans for the next step. Previously, China ranks among the first twenty-two countries to submit the Voluntary National Review (VNR) to the UN High-level Political Forum.

Poverty reduction is the primary goal set forth in the *2030 Agenda for Sustainable Development*, and targeted poverty reduction is also a major strategic deployment in China's implementation of the 2030 Sustainable Development Goals. China's Thirteenth Five-Year Plan put forward two "guarantees," namely to guarantee that all rural poor people are lifted out of poverty and ensure that all poor counties remain out of poverty. Accordingly, by 2020, all 70 million rural poverty people were lifted out of poverty; that is, more than 10 million people were lifted out of poverty every year. By the end of 2015, the State Council launched an investigation on poverty reduction, showing that of the 70 million poverty farmers in China, 42% were poor due to illness, 20% were poor due to disaster, 10% were poor due to education, 8% were poor due to the low labor capacity and 20% were poor due to other reasons. And the vast majority of these poverty farmers had no income increasing industry.[20] The *Decision of the CPC Central Committee and the State Council on Winning the Tough Battle against Poverty* adopted in November 2015[21] put forward "five measures" for targeted poverty reductions. According to the deployment of the central government, by 2020, 30 million people were lifted out of poverty through industrial support, 10 million were lifted out of poverty through transfer employment and 10 million were lifted out of poverty through relocation; the remaining more than 2,000 people were recognized as completely or partially incapable of work, all covered by subsistence allowances and lifted out of poverty through social security policy.

5.4. Establishment of China's innovation demonstration zones for the implementation of the 2030 Agenda for Sustainable Development

It is a common practice in China to mobilize the enthusiasm of local governments, and explore and summarize experience by the point-to-area method. In December 2016, the State Council issued the *Development Plan of China's Innovation Demonstration Zones for the Implementation of the 2030 Agenda for Sustainable Development*,[22] which planned to create about ten national innovation demonstration zones for the 2030 Agenda for Sustainable Development during the Thirteenth Five-Year Plan period. With a high enthusiasm in participation, a total of fifteen provincial (district) governments sent letters to the Ministry of Science and Technology to apply for building the demonstration areas. After field study, expert review and joint conference review, in March 2018, the State Council Information Office held a press conference on the construction of national innovation demonstration zones for the 2030 Agenda for Sustainable Development;[23] Xu Nanping, Vice Minister of the Ministry of Science and Technology, announced that the State Council officially approved Taiyuan, Guilin and

Shenzhen as the first batch of demonstration zones. The three cities were selected due to their representative regional characteristics; Shenzhen is a developed area, Taiyuan is a resource-based city, and Guilin is a tourist city, they are all very characteristic in different regional nature. Besides, all of the three places have a good work foundation. There were four main tasks to build the national innovation demonstration zones of the *2030 Agenda for Sustainable Development*. First, the sustainable development plan should be formulated based on the *2030 Agenda for Sustainable Development* and local demands. Second, bottleneck problems restricting sustainable development should be solved by strengthening technology screening, defining technology routes and then forming mature and effective systematic solutions. Third, the capacities of integrating and pooling innovative resources and promoting coordinated economic and social development should be enhanced to explore new mechanisms for integrating scientific and technological innovation and social development. Fourth, the experience in supporting sustainable development with scientific and technological innovation should be actively shared to play a radiating and driving role in other regions and provide Chinese approach for sustainable development to the world.

6. EXPERIENCE AND INSPIRATION OF CHINA'S HISTORY OF SUSTAINABLE DEVELOPMENT

China's history of sustainable development over the past forty years, especially the great changes in the past forty years of reform and opening-up, could be summarized in the following aspects.

First, China has constantly deepened the ideological understanding of sustainable development. Although ancient China had the philosophy of "harmony between man and nature," the modern concept of sustainable development originated from the West and was passively accepted in China at the very beginning. In the early stage, some local governments even used the term "continuous development" rather than "sustainable development" in documents, emphasizing that "development is the core of sustainable development." Along with rapid economic development after reform and opening-up, population, resource and environmental problems have become increasingly prominent, posing serious constraints on social and economic development and making it difficult to sustain the extensive growth pattern. During the profound reflection of realistic difficulties, the concept of sustainable development has gradually gained popularity among people. Sustainable development has been established as a national strategy, and the implementation of sustainable development strategy has gradually changed from passive acceptance to active choice, and has been promoted to a new height of ecological conservation.

Second, the thought on ecological civilization has contributed to the *2030 Agenda for Sustainable Development*. The *2030 Agenda* was formulated in a different way from the MDGs. The process was not only a consultation between politicians but also a great global discussion on the development and future destiny of mankind, attracting attention of the whole society through various channels. Development under the concept of industrial civilization has widened the development gaps between the North and the South, enlarged the disparity between the rich and the poor, and created serious inequities in allocation and consumption of environment and resources, with a lack of social justice. The *2030 Agenda for Sustainable Development*, which is themed on "Transforming Our World," calls for a profound reflection of industrial civilization and efforts in transforming the development paradigm, terminating poverty, realizing life transformation and earth protection and moving toward human dignity. This is not only a reflection and criticism of industrial civilization but also a pursuit of the transformation of development paradigm and the overall transformation of civilization.[24] The formulation and implementation of the *2030 Agenda for Sustainable Development* have almost kept pace with China's efforts in elevating ecological conservation to new heights, with highly consistent targets and content. Specifically, the *2030 Agenda* embodies the thought on ecological civilization, while the thought on ecological civilization provides the theoretical foundation for the implementation of the *2030 Agenda* and points the way forward.

Third, China has been repositioned along with the changes in global sustainable development pattern. Over the past half century, patterns of international geopolitics, world economic, population, energy and climate governance have all changed dramatically. The status of developed countries has declined, while that of emerging economies has risen. With the rapid economic growth since the reform and opening-up, China has witnessed a rising international status and gradually moved to the core of international governance. Now, China is undergoing an economic transition, covering population, economy, society, consumption, environment and other aspects, which is an overall transition from industrial civilization to ecological civilization. China's status in the world's pattern is different from that of other developing countries, emerging economies and developed countries. China needs to objectively understand the new changes and features of the international pattern of global sustainable development, and transforms its positioning based on its specific national conditions. Accordingly, China will turn from a regional power to a global power and shoulder more international responsibilities, which require both calm thinking and hard work.

Fourth, China has contributed Chinese wisdom and Chinese approach to global sustainable development. In the report to the Nineteenth CPC National

Congress, Chinese President Xi Jinping stated that China will be an important participant, contributor and pacesetter in global ecological conservation in the new era. China has transformed itself from a follower to an important participant, contributor and leader in the process of global sustainable development. The *2030 Agenda* not only aims to set goals but also embodies the process for countries to explore their respective development paths suitable for their national conditions. China's achievements in implementing the MDGs have attracted global attention. In recent years, there have been many good practices in poverty alleviation, haze control and ecological restoration, which need to be summarized and shared with other developing countries. As China is the largest developing country in the world, its development experience is of great exemplary significance in formulating and implementing the *Sustainable Development Goals for 2030*. With the rise of China's international status, the international community has higher expectations on China. In response, China proposed to build a community with a shared future for mankind and put forward the Belt and Road Initiative, with an aim to promote the implementation of the *Sustainable Development Goals for 2030* through the green Belt and Road. In a sense, this is a Chinese approach that China contributed proactively to global sustainable development.

7. CONCLUSION

In the past forty years of reform and opening-up, China has not only made great achievements in rapid economic growth, but also brought a series of severe challenges in population, resources and environment. Reflecting on the unsustainable development pattern, China has been exploring and practicing a sustainable development path suitable on its national conditions. While addressing its own problems, China has also made active contributions to global sustainable development. Besides, China has also shifted its roles from a spectator to a participant, contributor and leader.

Sustainable development, as a global consensus, is an eternal theme pursued by human society. With both opportunities and challenges in the future, the process of global sustainable development will be constantly deepened. At the National Conference on Ecological Environmental Protection, Chinese President Xi Jinping put forward new requirements of "working together on global ecological civilization construction and getting deeply involved in global environmental governance to come up with a worldwide solution for environmental protection and sustainable development, while guiding international cooperation to tackle climate change." In the new era, under the guidance of the thought on ecological conservation as well as the development concepts of innovation, coordination, green, openness and sharing, China will usher in a new chapter of sustainable development.

Notes

1 The Ten Measures for Environment and Development in China, *Environmental Engineering*, No. 2, 1993.
2 Report of the People's Republic of China on the Outline of the Ninth Five-Year Plan for National Economic and Social Development and the Long-Range Objectives through the Year 2010, http://www.npc.gov.cn/wxzl/gongbao/2001-01/02/content_5003506.htm.
3 Deng Nan (1994). On China's Agenda 21, *China Soft Science*, No. 10, 1994.
4 China's Agenda 21--White Paper on China's Population, Environment and Development in the 21st Century, 1994.
5 Path of Sustainable Development: Proceedings of the 1st Scientific Symposium on Sustainable Development at Peking University, Peking University Press, 1995.
6 Actions and Progress of Local Agenda 21 in China, *Environment and Development*, February 22, 2001, p. 3, website of Shanghai Environmental Protection Publicity and Education Center, http://www.envir.gov.cn/info/2001/2/222405.htm.
7 The United Nations: Millennium Development Goals Report 2015, http://mdgs.un.org/unsd/mdg/Resources/ Static/Products/Progress2015/Chinese2015.pdf.
8 The Ministry of Foreign Affairs of the People's Republic of China, United Nations in China: Report on China's Implementation of the Millennium Development Goals (2000–2015), http:// www.cn.undp.org/content/china/en/home/ library/mdg/mdgs-report-2015-/.
9 Transforming our World: The 2030 Agenda for Sustainable Development, http://www.un.org/ga/search/ view_doc.asp?symbol=A/RES/70/1&Lang=C.
10 The sustainable development goals could be roughly divided into four groups by content: goals 1–7: the goals involve eradicating poverty and hunger, guaranteeing the right to education, promoting gender equality and accessing water, sanitation and energy services, mainly reflect basic demands for guaranteeing human development, especially basic rights of vulnerable groups; goals 8–11: the goals involve sustainable economic growth and employment, sustainable industrialization and innovation, reducing inequality, building sustainable cities and human settlements, sustainable consumption and production, focus on promoting sustainable economic growth and social inclusion; goals 13–15: the goals involve addressing climate change, protecting marine resources and terrestrial ecosystems, emphasizing environmental sustainability; goals 16–17: the goals involve institutional construction, implementation means and partnerships, aim to strengthen the implementation of the goals through international cooperation.
11 *China's Position Paper on the Development Agenda beyond 2015* released by the Ministry of Foreign Affairs, website of Ministry of Foreign Affairs, http://www.gov.cn/gzdt/2013~09/22/content_2492606.htm.
12 Xi Jinping (2015). Working Together to Build New Partnership of Win-Win Cooperation and to Build a Community with a Shared Future for Mankind, Gmw.cn, September 29, 2015, http://news.gmw.cn/2015-09/29/content_17205547.htm.

13 Innovative Growth That Benefits All – Remarks on the World Economy At Session I of the 10th G20 Summit, November 16, 2015, http://news.gmw.cn/2015-11/16/content_17737908.htm.
14 China's Position Paper on the Implementation of the 2030 Agenda for Sustainable Development, website of the Ministry of Foreign Affairs, April 22, 2016, http:// www.fmprc.gov.cn/web/wjb_673085/zzjg_673183/gjjjs_674249/ xgxw_674251/t1356278.shtml.
15 The Communique of the fifth Plenary Session of the 18th CPC Central Committee, Caixin.com, October 29, 2015, http://www.caixin.com/2015-10-29/100867990.html.
16 Coordination Meeting on Implementation of the 2030 Agenda for Sustainable Development chaired by the Department of International Economic Affairs of the Ministry of Foreign Affairs, website of the Ministry of Foreign Affairs, November 5, 2015, http://www.fmprc.gov.cn/web/wjbxw_673019/t1312600.shtml.
17 Li Keqiang's Speech at the Symposium on the "Sustainable Development Goals: Working Together to Transform Our World--China's Proposal", website of the Ministry of Foreign Affairs, http://www.fmprc.gov.cn/web/ziliao_674904/zyjh_674906/t1399038.shtml.
18 China's National Plan on Implementation of the 2030 Agenda for Sustainable Development, website of the Ministry of Foreign Affairs, October 12, 2016, http://www.fmprc.gov.cn/web/zyxw/t1405173.shtml.
19 The China's Progress Report on Implementation of the 2030 Agenda for Sustainable Development, website of the Ministry of Foreign Affairs, August 2017, http://www.fmprc.gov.cn/web/ziliao_674904/zt_674979/dnzt_674981/qtzt/ 2030kcxfzyc_686343/P020170824649973281209.pdf.
20 Poverty Alleviation Office of the State Council: 42% of 70 Million Poor Farmers Due to Disease in China, News.163.com, December 16, 2015, http://news.163.com/15/1216/04/BAUAN08H0001121M.html.
21 Decision of the CPC Central Committee and the State Council on Winning the Tough Battle against Poverty, the government's website, December 7, 2015, http:// www.gov.cn/zhengce/2015-12/07/content_5020963.htm.
22 The Press Conference of the State Council on Issuing the Development Plan of China's Innovation Demonstration Zones for the Implementation of the 2030 Agenda for Sustainable Development (GF [2016] No. 69), website of the State Council, December 31, 2016, http://www.gov.cn/zhengce/content/2016-12/13/content_5147412.htm.
23 Press Conference held by the Information Office on the Development Plan of China's Innovation Demonstration Zones for the Implementation of the 2030 Agenda for Sustainable Development, website of the State Council, http:// www.gov.cn/xinwen/2018-03/23/content_5276861.htm#1.
24 In December 2014, UN Secretary-General Ban Ki-moon submitted the comprehensive report entitled "The Road to Dignity by 2030: Ending Poverty, Transforming All Lives and Protecting the Planet", http://www.un.org/en/ga/search/view_doc.asp?symbol=A/69/700&referer=http://www.un.org/millenniumgoals/&Lang=C.

CHAPTER EIGHT

Sustainable Urban Development: Practice and Experience

WANG MOU, KANG WENMEI, LIU JUNYAN, LV XIANHONG, ZHANG YING AND LUO DONGSHEN*

1. INTRODUCTION

During the forty years of reform and opening-up, along with the deepening of reform and opening-up and the rapid advancement of industrialization and urbanization, China's urban development has entered a new stage. From 1978 to 2017, the number of cities in China increased from 193 to 657 (in 2016);[1] the total permanent urban residents grew from 172.45 million to 813.47 million, and the proportion in China's total population rose from 17.9% to 58.52%.[2] On the one hand, urbanization gives birth to urban modernization, promotes the comprehensive development of economy, society and culture, and improves the

---

* Wang Mou, Doctor, Secretary-General and Associate Research Fellow of Research Center for Sustainable Development of Chinese Academy of Social Sciences, research fields: sustainable cities, regional development and environmental governance. Kang Wenmei, a postgraduate student of Chinese Academy of Social Sciences, research fields: econometric analysis and policy analysis of environmental and sustainable development issues. Liu Junyan, Doctor, Director of Climate and Energy Project of Greenpeace, research fields: ecological economics, ecosystem service accounting and sustainable development economics. Lv Xianhong, a lecturer of School of Humanities, Tianjin Agricultural University, research fields: sustainable development economics. Zhang Ying, Doctor, Associate Research Fellow, research fields: econometrics. Luo Dongshen, Doctor, Assistant Research Fellow, research fields: low-carbon economy and sustainable development.

living standard of urban residents; on the other hand, it exacerbates city-centered environmental pollution, health problems caused by environmental pollution and other urban problems. During the advancement of the reform, China has been exploring coordinated development of economy, society and environment. In September 1979, the Eleventh Session of the 5th National People's Congress adopted the *Environmental Protection Law of the People's Republic of China* (For Trial Implementation). In January 1984, the second National Environmental Protection Conference was held in Beijing, which established environmental protection as a basic state policy. In May 1984, the Environmental Protection Committee of the State Council was established, and environmental protection was formally incorporated into the national plan. Since the 1980s, as the concept of sustainable development gained global consensus, sustainable cities have increasingly become the focus of sustainable development research and practice. Since Habitat III was held in Quito on October 17–20, 2016, a global action framework for sustainable urban construction is being formed. Cities, as important carriers of social life, production and consumption in human society, have become the key battlefield of global sustainable development.[3]

During the continuous advance of reform and opening-up over the past forty years, along with the development of science and technology and the improvement of awareness of resources, environment and social governance in the human society, new concepts of environmental friendliness, resource conservation and equitable development have gradually moved from theory to practice, and guided the construction and development of sustainable cities in China. Over the past forty years, China has put forward and carried out a series of pilot and demonstration cities, including sustainable cities, ecological cities, eco-park cities, low-carbon cities, livable cities, resilient cities, sanitary cities, sponge cities and circular economy cities, making remarkable achievements and gaining development experience. This chapter reviews the overall process of sustainable urban construction and development in China, analyzes the development features of different stages, compares the features of different types of pilot and demonstration cities, and forecasts the future development prospects of sustainable cities.

2. OVERALL PROCESS OF SUSTAINABLE URBAN DEVELOPMENT

Sustainable development and sustainable urban construction have come along with China's reform and opening-up over the past forty years. China has actively participated in the process of sustainable development of the international community, and responded to and implemented the consensus and goals and tasks reached by the international community in terms of urban construction. After every important international summit, China has organized a series of practical activities on urban construction. In general, China's sustainable urban construction and development has gone through the initial stage of individual topics and individual construction and entered the exploration period focusing on comprehensive development and coordinated development. China's sustainable urban construction not only embodies the implementation of the global sustainable development agenda but also reflects the concrete practice of ecological conservation in the new era.

2.1. Promotion of China's sustainable urban construction in the process of global sustainable development

The concept of sustainable city originated from the theory of sustainable development. In 1984, the UN established the World Commission on Environment and Development; the latter issued a report entitled Our Common Future in 1987 and put forward the concept of sustainable development for the first time, and held global sustainable development summits in 1992, 2002 and 2015, in a bid to form a global consensus on sustainable development. China is an active participant to the conferences and practitioner of their achievements, and each conference has promoted the sustainable urban construction in China.

The UN Conference on Environment and Development was held in Rio de Janeiro, Brazil, from June 3 to 14, 1992, and approved such important documents as the *Rio Declaration on Environment and Development*, the *Agenda 21* and the *Statement of Principles on Forests*. Then Chinese Premier Li Peng led the Chinese delegation to the conference. After the Conference on Environment and Development, the Chinese government discussed and adopted *China's Agenda 21-White Paper on China's Population, Environment and Development in the 21st Century* at the 16th Executive Meeting of the State Council on March 25, 1994, putting forward China's goals of building sustainable cities, which feature a rational planning and layout, complete supporting facilities, convenient working and living conditions, clean, beautiful and quiet residential environment, and comfortable dwelling conditions.[4] China is one of the earliest countries to propose and implement sustainable development strategy as well as one of the earlier countries to start sustainable urban construction. In March 1996, the fourth Session of the Eighth National People's Congress approved the *Report of the People's Republic of China on the Ninth Five-Year Plan for National Economic and Social Development and the Long-Range Objectives to the Year 2010*, which listed sustainable development as an important national development strategy in the form of the state's supreme law for the first time and promoted the sustainable development in China and the process of sustainable development in urban areas. Subsequently, construction of sustainable cities, like sanitary cities, healthy cities, park cities and model cities for environmental protection, developed vigorously. In 1997, the Nineteenth Special Session of the UN General Assembly was convened to evaluate the progress in the implementation of *Agenda 21* over the past five years. State Councilor Song Jian led a delegation to the session and submitted *National Report on Sustainable Development of the People's Republic of China*. China's achievements in promoting sustainable development and sustainable urban construction were recognized by the international community.

The first World Summit on Sustainable Development (WSSD), held in Johannesburg, South Africa, from August 26 to September 4, 2002, was an important meeting for comprehensively reviewing and evaluating the implementation of *Agenda 21* and promoting the global partnership for sustainable development. It was subsequent to the UN Conference on Environment and Development in Rio de Janeiro, Brazil, in 1992, and the Nineteenth Special Session of the General Assembly in New York in 1997. In August 2002, the Chinese government issued the *Report on Sustainable Development of the People's Republic of China*. According to the report, the strategy is informed by a people-centered approach, a vision for harmony between man and nature, and the need for eco-friendly economic development. Science and technology and institutional reform will play a critical role in improving the well-being of the people. We must never falter in our efforts to promote the sustainable development of economy, society, population, resource and environment, while strengthening China's composite national strength and competitiveness, and carried out positive practice in cities. In 2002, the Seventeenth CPC National Congress clearly put forward the task of ecological conservation, and deployed and promoted the exploration of theoretical research and practical approach for ecological civilization. In the same year, the State Environmental Protection Administration launched the construction plan of eco-provinces, eco-cities and eco-counties. In 2004, the Ministry of Construction and the State Forestry Administration launched the construction of ecological park cities and forest cities, respectively. In 2008, China issued a series of policy documents on ecological protection, including the *National Ecological Function Zoning* and the *Outline of the National Key Ecological Function Areas*.[5] Low-carbon city pilot projects were carried out in 2010 and 2012, and "sponge cities" were studied and advanced since April 2012. In 2014, China adopted the *National New Urbanization Plan (2014–2020)*, which clearly stated that all cities shall achieve ecologically sustainable development, while significantly increasing public services and focusing on social equity.

In September 2015, 193 member states of the UN held the Sustainable Development Summit at the UN Headquarters in New York, where the *2030 Agenda for Sustainable Development* was formally adopted and implemented on January 1, 2016. The agenda included 17 sustainable development goals (SDGs) and 169 targets. Among them, the Eleventh SDG involved sustainable cities, which is stated as "make cities and human settlements inclusive, safe, resilient and sustainable."[6] In September 2016, China published the *National Plan on Implementation of the 2030 Agenda for Sustainable Development*, which not only defined the implementation of the *2030 Agenda for Sustainable Development* but also charted the path for China's future development. In December 2016, the State Council issued the *Development Plan of China's Innovation Demonstration Zones for the Implementation of the 2030 Agenda for Sustainable Development*, making clear arrangements for the construction of demonstration zones. By the end of April 2018, the State Council had officially approved three cities as the first batch of innovation demonstration zones on sustainable development, namely Taiyuan in Shanxi Province, Guilin in Guangxi Zhuang Autonomous Region and Shenzhen in Guangdong Province. Pilot projects for low-carbon city, climate-adaptive cities, sponge cities, resilient cities, national comprehensive new-type urbanization and pilot zones for national comprehensive supporting reform were also established to advance the construction of sustainable cities in different aspects and implement various sustainable development goals on the basis of *China's National Plan on Implementation of the 2030 Agenda for Sustainable Development*.

2.2. Practice of sustainable cities in China

The process of sustainable urban development in China could be roughly divided into three stages by time series. The first stage is from 1986 to 2000, which is the initial practice stage of sustainable urban development; the second stage is from 2001 to 2012, focusing on the transformation and development during sustainable urban construction; the third stage is from 2012 to present, emphasizing coordinated development and paying attention to social equity during sustainable urban construction. The stages of sustainable urban development roughly correspond to China's economic development stages, levels and features. In the first stage, China witnessed a relatively low economic and social development level, while the international community was also in the initial stage of the sustainable development. Therefore, sustainable urban construction in this stage mainly focused on special problems, especially those affecting the subsistence of urban residents. It could be seen that in the 1990s, China carried out a series of pilot and demonstration projects in terms of sanitation, infrastructure, environment, health and landscape. These individual practices led by relevant ministries and commissions also contributed experience and laid a foundation for the subsequent comprehensive practice of sustainable cities (as shown in Figure 8-1).

China joined the World Trade Organization in 2001. With the vigorous development of the foreign trade processing industry, China saw an accelerated overall industrialization and then a high-speed urbanization. Cities became the main carrier of industrial cluster and population aggregation; in turn, cities have developed rapidly, thanks to the industrial cluster and population aggregation. However, the extensive urban expansion model without constraints and top-level design was soon challenged, as evidenced by the fact that insufficient urban resources, environmental pollution, traffic congestion, poor air quality and other problems appeared intensively, and the urban transformation and development pattern became a social concern. At the city level, demonstration projects for the transformation and upgrading of old industrial cities and resource-dependent cities were carried out under the leadership of the National Development and Reform Commission, the Ministry of Science and Technology, and the Ministry of Industry and Information Technology; low-carbon development pilot cities and circular economy cities led by the National Development and Reform Commission, and sponge cities and smart cities led by the Ministry of Housing and Urban-Rural Development became sustainable urban practices to explore the economic and social transformation and development pattern.

Since 2012, China's economic and social development has entered a new stage, in which qualitative improvement has played an increasingly important role in the game between quantitative expansion and qualitative improvement. The new government has clearly put forward the "five-in-one" goal of harmonious development in economic construction, political construction, cultural construction, social construction and ecological conservation, fully covering the three dimensions of sustainable development. Urban practice has also paid more attention to comprehensiveness and integrity, focused on the top-level design of urban development, and attached great importance to social equity. Sustainable urban construction practices during this period included the national comprehensive new-type urbanization pilot projects, focusing on integration of migrant workers into cities, cultivation of new small and medium-sized cities, green and intelligent development of cities (towns), integrated development of industries and cities and other equitable and coordinated development issues; innovation demonstration zones on sustainable development, aiming to solve principal challenges facing China in the new era, form realistic models and typical patterns for sustainable development innovation demonstration, play exemplary and leading roles in sustainable development for other regions in China and provide China's experience in implementing the *2030 Agenda for Sustainable Development* for other countries; and national experimental zones for integrated, complete reform, with the purpose of exploring new patterns of regional development under new historical conditions and promoting reforms in key areas, such as innovation-driven development, integrated urban-rural development, all-round opening-up and green development.

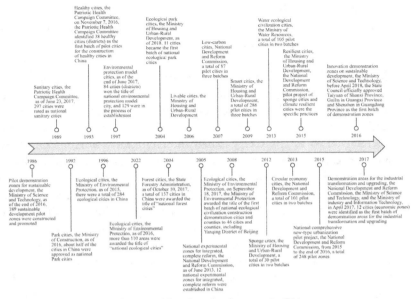

Figure 8-1 Overall process of sustainable urban development in China

3. MAIN PRACTICES AND FEATURES OF SUSTAINABLE URBAN CONSTRUCTION

During the deepening of reform and opening-up, according to the requirements and features of sustainable development in different stages, China has proposed several plans to establish sustainable development pilot zones, ecological cities, national experimental zones for integrated, complete reform, low-carbon cities, sponge cities, innovation demonstration zones on sustainable development, and demonstration projects for the transformation and upgrading of old industrial cities and resource-dependent cities in aspects of society, economy and environment. Each concept in the field of sustainable urban development has a specific historical background, target positioning, connotation, and focus. Therefore, this chapter briefs on and compares the idea and features, and practice and development of the above-mentioned urban concepts based on the time when they were put forward in China.

3.1. Sector-based pilot and demonstration projects featuring improvement of living conditions

This was the main way of promoting sustainable cities in the early stages in China. With a relatively low accumulation of social wealth, China mainly focused on survival-related issues, such as environment, sanitation and health, which was consistent with the features of economic and social development at that time and laid a foundation for subsequent development.

(1) Sanitary cities: In order to promote urban sanitary infrastructure construction in China, strengthen urban sanitary management, improve urban sanitary conditions, promote economic development and enhance sanitary awareness, the Patriotic Health Campaign Committee (hereafter referred to as "PHCC") issued the *Notification of Establish National Sanitary Cities* in 1989. Subsequently, national sanitary cities were established in full swing. From the point of view of environmental sanitation construction, PHCC proposed the National Urban Sanitation Standards and the Measures for Assessing and Naming National Sanitary Cities.[7] By June 23, 2017, 297 cities had been rated as national sanitary cities nationwide.[8]

(2) Healthy cities: The concept of "Healthy cities" is closely related to sustainable cities, and was proposed in the 1980s based on the "New Public Health Movement," the *Ottawa Charter* and the goal of "health for all and all for health." In 1986, the World Health Organization (WHO) first adopted the concept of a healthy city in the Initiatives on Healthy Cities and Villages and put forward the official definition of a healthy city in 1992. In August 1994, with the help of the WHO, healthy city pilot projects were launched in Dongcheng District of Beijing and Jiading District of Shanghai, marking the official launch of healthy urban construction in China. On June 12, 2001, the Patriotic Health Campaign Committee Office (hereafter referred to as "PHCCO") officially declared Suzhou as the first pilot "healthy city" in China to WHO. At the end of 2007, PHCCO designated ten cities (districts and towns), including Shanghai, as the first batch of pilot healthy cities in China, opening a new chapter in China's healthy urban construction. In 2012, the State Council approved the implementation of the Twelfth Five-Year Plan for development of public health, and officially announced the launch of a campaign to build healthy cities and towns.[9] On October 25, 2016, the CPC Central Committee and the State Council released the *Outline of the Healthy China 2030 Plan*. On November 7, 2016, PHCCO designated thirty-eight national sanitary cities (districts) as the first batch of pilot cities of healthy urban construction nationwide. On April 9, 2018, PHCC published the National Evaluation Index System for Healthy Cities, which was closely related to the goals and tasks of healthy urban construction in China, and focused on health problems

and factors influencing health in China's urban development in the current stage.[10]

(3) Park cities: National park cities were established by the Ministry of Construction in 1992. In 1992 and 1996, 12 "selection standards for Park cities" were formulated and revised. In May 2000, the *Implementation Plan for Establishment of National Park Cities* was formulated.[11] In 2005, the Ministry of Construction decided to designate thirty-one cities, including Wuhan, as "national park cities." Over the past twenty years, China has made remarkable achievements in urban landscaping, with a substantial increase in the total green land area by 4.7 times in cities nationwide. More than 60% of the cities have established green line control and system of keeping the public informed, and most of the cities had the park green land service radius coverage close to or exceeding 80%.[12] With the acceleration of China's urbanization, the contradiction between the backward urban infrastructure construction and the improved living environment of people has become increasingly prominent. In the context that landscaping has gradually played a basic role in improving urban ecology and living environment, the "upgraded version" of national park cities have come into being. In 2004, the Ministry of Construction started to establish national eco-park cities. In 2006, Shenzhen became the first national eco-park demonstration city. In 2007, the Ministry of Construction selected 11 cities, including Qingdao, as the pilot cities. In 2012, the Grading and Assessment Standards for National Ecological Park Cities were issued to define comprehensive assessment methods involving remote-sensing testing and field expert study. On January 29, 2016, seven cities, including Xuzhou, became the first batch of national ecological park cities.[13] On October 27, 2017, the Ministry of Housing and Urban-Rural Development decided to designate four cities, including Hangzhou, as national eco-park cities.

(4) Forest cities: China's urban forest construction began in the 1980s. With the active promotion and support from the government, forest urban construction has been booming around China. On November 18, 2004, the first China Urban Forest Forum was held in Guiyang, Guizhou Province, which was approved as China's first national forest city, marking the official ushering of China's national forest urban construction. On March 15, 2007, the State Forestry Administration announced the national forest city assessment indexes. In 2013, the State Forestry Administration compiled the *Outline on Promoting Ecological Conservation Plan* (2013–2020), proposing to "vigorously carry out forest urban construction activities." On January 26, 2016, Chinese President Xi Jinping stressed to vigorously carry out forest urban construction at the Twelfth meeting of

the CPC Central Leading Group for Financial and Economic Affairs. In March 2016, the Thirteenth Five-Year Plan clearly proposed to support forest urban construction, making forest urban construction a national strategy. In August 2016, the State Forestry Administration published the Measures for Approving the Title of National Forest Cities (Exposure Draft). The Thirteenth Five-Year Plan for Forestry Development clearly put forward to build six national forest city clusters in Beijing, Tianjin and Hebei by 2020.[14] As of October 10, 2017, a total of 137 cities across China were awarded the title of "national forest cities."

(5) Environmental protection model cities: The concept of national environmental protection model cities was proposed by the State Environmental Protection Administration in accordance with the Nineth Five-Year Plan for National Environmental Protection and the Vision for 2010, covering society economy, environment, urban construction, health, landscape and other aspects. It involved a wide range of areas, with a high starting point and difficulty in implementation, and only the national sanitary cities that pass the quantitative assessment for comprehensive urban environment renovation and reach a certain standard for environmental protection investment are qualified for application.[15] At the 4th National Conference on Environmental Protection in 1996, the State Environmental Protection Administration awarded Zhangjiagang the title of "National Environmental Protection Model City," marking the formal start of the model city establishment work. In 1997, the State Environmental Protection Administration issued the *Notification on Carrying out Activities for Establishing National Environmental Protection Model Cities*, which began to standardize and institutionalize the model city establishment work. Subsequently, relevant administrative measures were adjusted for three times in total. In 2008, the Administrative Measures for Establishing National Environmental Protection Model Cities was issued to define the complete procedures and system from plan development and implementation, application submission, recommendation by provincial departments, technical evaluation, assessment and acceptance inspection, public announcement and five-year review, marking the gradual maturity of the model city establishment management system. In 2011, the National Environmental Protection Model City Assessment Indexes and the Detailed Rules for Implementation (Phase VI) were released, which paid more attention to environmental quality and emphasize pollution and emission reduction in addition to the previous index system. The dynamic adjustment of the index system advanced the continuous improvement and progress of environmental protection model cities.[16]

The title of environmental protection model city is valid for five years rather than for long. As of June 30, 2017, eighty-four cities (areas) were awarded the title of national environmental protection model cities, and 129 cities (areas) were being established, covering thirty provinces.[17]

3.2. Pilot and demonstration projects for promoting transformation and development

With the rapid advance of industrialization and urbanization in China, the rapid urban expansion pattern featuring industrial cluster and population aggregation has been challenged by urban environment pollution, resources scarcity in urban areas, insufficient urban growth momentum and declined quality of urban life. Explorations shall be made for an urban sustainable development pattern adapted to the new development stage, involving key industry development transformation, low-carbon development transformation and smart development transformation in cities.

(1) demonstration zones for the transformation and upgrading of old industrial cities and resource-dependent cities: old industrial cities and resource-dependent cities had risen due to industrial development and resource exploitation. However, problems have gradually come along as economic development entered a new normal and turned to high-quality development. For example, these cities feature a large proportion of heavy industries and traditional industries in the industrial structure, a narrow industrial chain extension scope and insufficient scale effect, in addition to such problems as a weak decisive role of markets, low efficiency of state-owned enterprises and underdeveloped private economy.[18] In order to promote the transformation and upgrading of old industrial cities and resource-dependent cities, the central government issued the *Plan for Adjusting and Renovating the Old Industrial Bases Throughout the Country (2013–2022)* and the *Sustainable Development Plan of Resource-dependent Cities Nationwide* (2013–2020), involving 120 and 242 cities, respectively. In April 2017, the National Development and Reform Commission, the Ministry of Science and Technology, the Ministry of Industry and Information Technology, the Ministry of Land and Resources, and China Development Bank jointly issued a notification, identifying twelve cities (economic zones) as the first batch of demonstration areas for the industrial transformation and upgrading, defining the specific tasks of the demonstration areas for the industrial transformation and upgrading, and preliminarily determining the supportive policies for the demonstration zones in terms of industry, innovation, investment, finance, land and

other aspects,[19] so as to guide sustainable development of old industrial cities and resource-dependent cities.

(2) Low-carbon cities: The concept of low carbon originally appeared in the field of economic development in the human society. In 2003, the UK first formally proposed the concept of a "low-carbon economy" in the *Energy White Paper*. Subsequently, the concept of low carbon extended from the field of economic development to the field of social life. In general, there are two common points in the interpretation of low-carbon cities. First, low-carbon cities achieve the same quality of economic development with lower carbon emissions under the premise of economic development; second, low-carbon cities are built in all aspects, including society, economy, environment and culture. China began to explore low-carbon urban construction in 2008, and the National Development and Reform Commission designated three batches of low-carbon pilot provinces and cities in 2010, 2012 and 2017, respectively. So far, there is at least one low-carbon pilot city in every province or autonomous region in China. Besides, there are also various forms of pilot and demonstration projects in cities in China, such as low-carbon pilot industrial parks, low-carbon pilot communities and low-carbon pilot towns.

(3) Smart cities: In 1998, the concept of "digital earth" was proposed, and then the concept of "digital city" came with it. In 2008, IBM proposed the smart earth; and in 2009, it put forward the slogan that the construction of "smart cities" is the first step for the construction of the "smart earth," hoping to lead the world's cities to prosperity and sustainable development through the construction of "smart cities." Smart cities are an upgraded version of digital cities. In 2009, China introduced the concept of smart urban construction.[20] In the Twelfth Five-Year Plan, more than twenty cities, including Beijing and Tianjin, regarded smart cities as their construction goals.[21] On December 5, 2012, the Ministry of Housing and Urban-Rural Development issued the Notification on Launching National Smart City Pilot Projects, marking the official launch of smart city pilot projects in China. In 2012, the *Ministry of Housing and Urban-Rural Development published the National Smart City (District, Town) Pilot Index System (for Trial Implementation)*, which was an important reference for establishing smart cities. In November 2013, the Ministry of Science and Technology and the Standardization Administration identified twenty cities as national "smart city" technology and standard pilot cities. On August 29, 2014, eight ministries and commissions, including the National Development and Reform Commission, issued the *Guiding Opinions on Promoting Healthy Development*

of Smart Cities, proposing to build a batch of smart cities with distinctive characteristics by 2020.[22] In March 2016, the Thirteenth Five-Year Plan proposed to strengthen modern information infrastructure construction, promote development of big data (and AI) and the Internet of things, and build smart cities. Since 2013, the Ministry of Housing and Urban-Rural Development has designated 286 cities (towns) as smart pilot cities in three batches, most of which are small units at the district, county, or even street or town level. After years of construction and development, smart urban construction has been in full swing in China and listed as one of the 100 key construction projects in the Thirteenth Five-Year Plan.

(4) Sponge cities: In April 2012, the concept of "sponge cities" was first proposed at the "2012 Low-Carbon City and Regional Development Technology Forum."[23] In recent years, with frequent occurrence of urban flood disasters, "sponge cities" and the corresponding planning concepts and methods have been recognized by all walks of life. According to the *Technical Guide for Building Sponge Cities—Rainwater System Construction for Low Impact Development (for Trial implementation)* issued by the Ministry of Housing and Urban-Rural Development, sponge cities were defined as the cities that are resilient as sponges in adapting to environmental changes and coping with natural disasters, and could absorb, store, permeate and purify water when it rains and "release" the stored water and use it when necessary, with the aim to improve urban ecosystem functions and reduce the occurrence of urban floods. On December 12, 2013, Chinese President Xi Jinping emphasized in his speech at the Central Urbanization Work Conference that "in improvement of the urban drainage system, priority should be given to reserving the limited rainwater, drain water with natural force and build sponge cities with natural storage, natural permeation and natural purification." In February 2014, the *Main Work Points of the Urban Construction Department of the Ministry of Housing and Urban-Rural Development* explicitly proposed to "urge local governments to accelerate transformation of rainwater and sewage water diversion, improve the urban drainage and waterlogging prevention level, vigorously promote the low-impact development and construction pattern, and speed up the study of policies and measures for building sponge cities." In March of the same year, Chinese President Xi Jinping stressed to "build sponge homes and sponge cities" again at the 5th meeting of the CPC Central Leading Group for Financial and Economic Affairs. On October 22, 2014, the Ministry of Housing and Urban-Rural Development issued the Technical Guidelines for Sponge City Construction.[24] In December of the same year, the Ministry of Finance, the Ministry of Housing and Urban-Rural Development and the

Ministry of Water Resources jointly issued the Notification on launching Sponge Urban Construction Pilot Work under Support of the Central Finance, and organize to launch the sponge city pilot and demonstration projects. In October 2015, the General Office of the State Council issued the *Guiding Opinions on Promoting Construction of Sponge Cities*, which required to permeate, retain, store, purify, use and drain 70% of rainfall for local recycling.[25] By 2020, more than 20% of urban built-up areas met the goals; by 2030, more than 80% of urban built-up areas will meet the goals. In 2017, Chinese Premier Li Keqiang in the Report on the Work of the Government defined the development direction of sponge citie; that is to say, sponge urban construction is not only limited to pilot cities, but all cities should pay attention to this "lining project."[26] Since April 2015, China has announced two batches of sponge urban construction pilot projects supported by the central finance, including sixteen sponge pilot cities in the first batch and fourteen sponge pilot cities in the second batch.

(5) Resilient cities: In recent years, serious problems in flood control, public security, air pollution and snowstorm have been exposed in Chinese cities, such as July 2012 flood in Beijing (July 2012), haze in Beijing, Tianjin, Hebei and Yangtze River Delta (December 2013), the biggest snowstorm in Yanqing District of Beijing in fifty-two years (November 2012), and massive flood in Wuhan (July 2016). They were beyond the coping capacity of the cities, and caused huge property losses and casualties, exposing the high vulnerability and low resilience of Chinese megacities. In response, we urgently need to understand the correlative mechanism between urban development and disaster risk, strengthen forward-looking planning, and improve the climate resilience of cities. In August 2011, the 2nd World Cities Scientific Development Forum and the 1st Mayors' Summit on Disaster Prevention and Mitigation were held in Chengdu; ten cities, including Chengdu, joined the plan for "making the cities more resilient," and discussed and approved the *Declaration of Chengdu Action of "Top 10" Index System for Making the Cities More Resilient* and the *Action Plan for Sustainable Urban Development*. In November 2013, the *National Climate Change Resilience Strategy* was released to identify urbanized areas as key resilient areas. At the end of 2015, the Central Urban Work Conference put urban security first, and placed the urban disaster prevention and resistance capacity at its core. In February 2016, the National Development and Reform Commission and the Ministry of Housing and Urban-Rural Development jointly issued the *Urban Climate Change Resilience Action Plan*, which proposed to build thirty climate-adaptive pilot cities based on the goals of "safety, livability,

green, health and sustainability" by 2030. So far, twenty-eight urban pilot zones have been announced.

(6) Circular economy cities: The concept of circular economy is based on repeated and economical utilization of resources, with the aim to maximize benefits in economic development with the least factor input and the least pollutant outflow.[27] In 1994, *China's Agenda 21* put forward the sustainable development goal of "promoting environmentally sound technology and developing circular economy," which started the process of building cities with circular economy. The government wrote the development of circular economy into its Eleventh Five-Year Plan as a national strategy. The government established the first batch of eighty-four circular economy demonstration zones in 2005 and the second batch of sixty-five circular economy demonstration zones in 2007. After the promulgation of the *Circular Economy Promotion Law* in 2008, China's circular economy has been developed in an all-round way.[28] In September 2013, the National Development and Reform Commission issued the *Notification on Organizing and Establishing Circular Economy Demonstration Cities (Counties)*, launching the establishment of circular economy demonstration cities (counties). The formal start of this work indicated that the development of circular economy in China gradually extended from "points" and "lines" to "areas" and from "industries" to the "society."[29] In December 2013, the National Development and Reform Commission officially identified forty areas, including Yanqing County of Beijing, as the national circular economy demonstration cities (counties) in 2013.[30] On September 24, 2015, the National Development and Reform Commission and other organizations compiled the Guidelines on Compiling Construction Implementation Plans for Circular Economy Demonstration cities (counties).[31] On January 6, 2016, the National Development and Reform Commission approved to designate sixty-one areas, including Jinghai District of Tianjin, as national circular economy demonstration cities (counties) in 2015.[32] In 2017, the Nineteenth CPC National Congress put forward new requirements for the development of circular economy, which charted the course of the development of circular economy and injected the impetus.

3.3. Comprehensive pilot and demonstration projects featuring coordination and equity

(1) Sustainable development pilot area: In 1984, the UN established the World Commission on Environment and Development, which issued the report entitled "Our Common Future" in 1987 and put forward the

concept of sustainable development. According to domestic development during the same period, driven by the reform and opening-up and the scientific and technological progress, the national economy began to recover, with rapid economic development in many areas, especially the eastern coastal areas. Meanwhile, however, there emerged such problems as backward social development and deteriorated ecological environment. In response to backward social development, deteriorated ecological environment and rising international environmental protection movements, in 1986, the State Science and Technology Commission and relevant ministries and commissions of the State Council started the comprehensive urban social development demonstration and pilot project in Changzhou and Huazhuang Town of Xishan District, Jiangsu Province, official launching the construction of pilot zones.[33] During the five years of the technology-guided comprehensive social development pilot project, "proactive achievements were made in mobilizing the positive factors of people, playing the role of science and technology as the primary productive force, and promoting coordinated and sound development of economy and society."[34] To this end, the State Science and Technology Commission and the State Commission for Restructuring Economy decided to build "comprehensive social development experimental zones" based on the comprehensive pilot work in 1992. In March 1994, the work center of experimental zones turned to sustainable development, and each experimental zone was required to take the lead in building a base for implementing *China's Agenda 21* and sustainable development strategy. Till 1997, there were up to twenty-eight comprehensive social development experimental zones. They were mainly divided into three types, namely urban areas, counties and towns in big cities, involving experiments focusing on urban and rural overall planning, social security, population health and garbage disposal. In 1996, China clearly put forward the strategy of rejuvenating the country through science and education and the strategy of sustainable development. In order to promote the implementation of the strategy of sustainable development and the strategy of rejuvenating the country through science and education, the "comprehensive social development experimental zones" was renamed as the "sustainable development experimental zones." Currently, the work of the experimental zones is promoted by an inter-ministerial conference composed of twenty departments. As of 2016, a total of 189 experimental zones had been built upon approval and distributed in thirty-one provinces, autonomous regions and municipalities. The ratio of experimental zones in eastern, central and western areas was 5:3:2, while the proportions of town, urban, prefecture and township-level experimental zones were

48%, 34%, 15% and 3%, respectively. The experimental themes covered various fields of sustainable development, such as economic transformation, social governance and environmental protection. A total of about 300 provincial-level sustainable development experimental zones were established in various provinces.

(2) Innovation demonstration zones on sustainable development: In order to implement the UN *2030 Agenda for Sustainable Development*, the State Council issued the *Program for Building the Innovation Demonstration Zones for China's Implementing the 2030 Agenda for Sustainable Development* in December 2016, with the aim to make clear arrangements for the construction of demonstration zones. In general, the plan was positioned to explore systematic solutions to sustainable development issues with science and technology at its core, set an example and play a leading role in China's efforts to solve major social contradictions and fulfill development tasks in the new era, and provide China's experience for global sustainable development, under the guidance of Xi Jinping Thought on Socialism with Chinese Characteristics for a New Era, in the principle of "innovative concept, problem-oriented approach, diversified participation, and open and shared benefits," and with the deep integration of scientific and technological innovation with social development as the focus. It mainly aimed to establish about ten national sustainable development agenda innovation demonstration zones during the Thirteenth Five-Year Plan period, form a number of realistic models and typical patterns for sustainable development innovation demonstration, play exemplary and demonstration roles in sustainable development for other zones in China, and provide China's experience to other countries in implementing the *2030 Agenda for Sustainable Development*.[35] As of April 2018, the State Council officially approved Taiyuan in Shanxi Province, Guilin in Guangxi Zhuang Autonomous Region and Shenzhen in Guangdong Province as the first batch of demonstration zones.

(3) National comprehensive new-type urbanization pilot projects: The conventional extensive urbanization pattern will bring many risks, such as slow industrial upgrading, deteriorated resources and environment, and increasing social contradictions, and may fall into the "middle-income trap," which will further affect the modernization process. Along with profound changes in internal and external environments and conditions, urbanization has to enter a new stage of transformation and development focusing on quality improvement. In 2014, the National Development and Reform Commission issued the Overall Implementation Plan of National Comprehensive New-Type Urbanization Pilot Project, which defined the overall requirements for various pilot tasks, focusing

on integration of migrant workers into cities, cultivation of new-born small and medium-sized cities, green and intelligent development of cities (towns), integrated development of industries and cities, transformation of development zones, redevelopment and utilization of inefficient urban land, coordinated development mechanism of city clusters and construction of new countryside. The National Development and Reform Commission announced a total of 248 national comprehensive new-type urbanization pilot zones in three batches in February 2015, November 2015 and December 2016, respectively.

(4) National experimental zones for integrated, complete reform: Along with rapid development of urbanization and industrialization processes, China's economic and social development is facing many new challenges, and its development and reform have entered a critical stage to adapt to the demands of rapid economic and social development. On June 21, 2005, Chinese Premier Wen Jiabao presided over an executive meeting of the State Council, which approved the comprehensive supporting reform pilot project for socialist market economy in Pudong New Area. According to the overall layout of China's regional economic development, explorations were made on the new regional development pattern under new historical conditions, opening a new chapter of the construction of national experimental zones for integrated, complete reform. Since 2005, the government has officially approved to establish twelve national experimental zones of integrated, complete reform, which could be divided into six categories by nature: (1) comprehensive supporting reform pilot zones, including Shanghai Pudong New Area, Tianjin Binhai New Area, Shenzhen, Xiamen and Yiwu; (2) comprehensive overall urban and rural development supporting reform pilot zones, including Chongqing and Chengdu; (3) comprehensive resource-conserving and environment-friendly society construction supporting reform pilot zones (hereinafter referred to as the comprehensive supporting reform pilot zones for two-oriented society), including Wuhan Metropolitan Circle and Changsha-Zhuzhou-Xiangtan Urban Agglomeration; (4) comprehensive new-type industrialization supporting reform pilot zones, including Shenyang Economic Zone; (5) comprehensive modern agriculturalization supporting reform pilot zones, including "two great plains" in Heilongjiang Province; (6) comprehensive resource-based economy supporting reform pilot zones, including Shanxi Province. In May 2018, the National Development and Reform Commission issued the Key Tasks of National experimental zones for integrated, complete reform in 2018, which focused on deepening reform in such key areas

as innovation-driven development, urban-rural integration, all-round opening-up and green development, and defined the key reform tasks of the twelve comprehensive supporting reform pilot zones in the new stage.[36]

(5) Ecological civilization cities: The concept of ecological civilization was first proposed by the agricultural economist Ye Qianji in 1984. The discussions about ecological civilization and the studies of ecological economics in China since the 1980s actually was in response to ecological degradation rather than environmental pollution and resource depletion, especially fossil energy. In the twenty-first century, China's large-scale rapid industrialization already exceeded the carrying capacity of resources and environment. The large-area haze, pollution of rivers and lakes, and heavy metal poisoning of soil in China[37] indicate qualitative changes in ecological imbalance. Accordingly, ecological civilization was put on the agenda again and placed in a prominent position. In 2007, the Seventeenth CPC National Congress clearly put forward the construction task of ecological civilization. In April 2015, the CPC Central Committee and the State Council adopted the *Opinions on Accelerating the Advancement of Ecological Civilization*, which was the first document of the central government to make special arrangements for ecological civilization construction and define the overall requirements, goals and visions, key tasks and institutional system of ecological civilization construction. On September 11, 2015, the meeting of the Political Bureau of the CPC Central Committee approved the *Integrated Reform Plan for Promoting Ecological Progress*, which was the top-level design and deployment of ecological civilization reform. On October 28, 2016, the Ministry of Environmental Protection issued the *Outline of the Thirteenth Five-Year Plan for National Ecological Protection*. The main goal was to build 60–100 ecological conservation demonstration zones and a number of environmental protection model cities, achieving an obvious demonstration effect for ecological conservation.[38] On September 18, 2017, the Ministry of Environmental Protection awarded forty-six cities and counties, including Yanqing District of Beijing, the title of the first batch of national ecological civilization demonstration cities (counties).

(6) Livable cities: In 1996, the 2nd UN Conference on Human Settlements put forward the concept that cities should be livable human settlements. Soon this concept reached a broad consensus in the international community and became a new view of city in the twenty-first century. In 2005, the concept of "livable cities" first appeared in Beijing's Urban Master Plan approved by the State Council. In 2007, the Department of Science and Technology of the Ministry of Construction issued the

Scientific Evaluation Standard for Livable Cities, which proposed the guiding scientific standards for building livable cities in six aspects, such as social civilization and economic prosperity, as China's first official and authoritative standards for building livable cities. In 2014, the *National New Urbanization Plan* (2014–2020) defined the construction of harmonious and livable cities as one of the important goals of the new-type urbanization. In 2015, the Central Urban Work Conference proposed to "vigorously solve prominent problems such as urban maladies, constantly improve the quality of urban environment, quality of people's life and urban competitiveness, and build harmonious and livable modern cities with vitality and distinctive characteristics." The Outline of the Thirteenth Five-Year Plan released in 2016 also specially stated to build harmonious and livable cities. At present, many cities in China have clearly taken "livability" as the goal of urban construction and included it in the urban master planning and space planning.

(7) Eco-cities: Destruction of natural resources and ecological environment in China is increasingly serious, and so in order to solve these problems, in 1995, the State Environmental Protection Administration issued the *Outline of Construction Plan for National Ecological Demonstration Zones* (1996–2050), so as to organize and launch the construction of ecological demonstration zones. The main features of eco-cities are as follows. First, they should have a sound ecological environment and move toward a higher level of balance, with almost no environmental pollution, and effective protection and rational utilization of natural resources. Second, a stable and reliable ecological security system has basically taken shape. Third, environmental protection laws, regulations and systems have been implemented effectively. Fourth, social and economic development featuring circular economy has been accelerated. Fifth, the cities feature the coexistence between man and nature in harmony and the significant development of ecological culture. Sixth, urban and rural environments are clean and beautiful, with improved living standard of people in an all-round way. On May 23, 2003, the State Environmental Protection Administration issued the *Construction Indexes for Eco-Counties, Eco-Cities and Eco-Provinces (for Trial Implementation)*. From June 2 to October 8, 2006, the State Environmental Protection Administration designated 183 national eco-cities (counties or districts) in 9 batches, including 13 eco-cities, 38 eco-counties (county-level cities), 70 eco-counties and 62 eco-districts.[39]

(8) Water ecological civilization cities: Since the end of 1990s, China has carried out a series of most direct theoretical and practical explorations, such as water-conserving society, the strictest water resource management

system, and river and lake health assessment, involving urban flood control and efficient utilization of water resources.[40] In 2013, the Ministry of Water Resources launched the construction of water ecological civilization cities in accordance with the deployment requirements of the CPC Central Committee on ecological conservation. In 2013 and 2014, 105 cities (counties or districts) were designated as national water ecological conservation pilot cities. By the end of December 2017, the first batch of forty-six water ecological civilization pilot cities had been completed and made remarkable achievements. The second batch of fifty-nine pilot cities was under construction and expected to be completed in 2018.[41]

(9) Green cities: The concept of "green cities" could be traced back to modernist master Le Corbusier's plan of "green cities" exhibited in Brussels in 1930. Sustainable development, green low-carbon development and other theories have promoted the practice of green cities. The book entitled *Green Cities*, which was edited by David Gordon, systematically defines the concept, connotations and implementation path of green cities. The concept of green cities was proposed in green movements for protecting global environment. For the purpose of coordinated development between nature and human, green cities not only emphasize ecological balance and natural protection but also focus on human health and cultural development.[42] China put forward the concept of "urban landscaping" in the Outline of the Tenth Five-Year Plan and the requirement of building "green cities" in the *Outline of the Thirteenth Five-Year Plan*. As the vast majority of provinces and municipalities have written green development into their local Thirteenth Five-Year Plan, green urban construction is being implemented through green industrial transformation and development, green transport, green consumption and other plans.

(10) Green ecological demonstration cities: Green ecological demonstration cities were proposed in the process of using local renewable energy and resources in new urban areas and promoting large-scale development of green buildings. In March 2013, the Ministry of Housing and Urban-Rural Development issued the Twelfth Five-Year Plan for the Development of Green Buildings and Green Ecological Urban Areas, proposing to implement 100 demonstration projects of green ecological urban areas by the end of the Twelfth Five-Year Plan period. In 2012, the Ministry of Housing and Urban-Rural Development designated the first batch of eight green ecological demonstration cities, including Sino-Singapore Tianjin Eco-City. In 2013, five green ecological demonstration zones, including Zhuozhou Ecological Livable Demonstration Base, were established. In 2014, six green ecological demonstration zones, including Changxindian Ecological Zone in Beijing, were established.

4. SUSTAINABLE URBAN DEVELOPMENT: EXPERIENCE AND PROSPECTS

Throughout the development history of sustainable cities in China during the forty years of reform and opening-up, it can be found that the focus of various experimental and pilot zones has shifted from individual construction to overall construction, and from the environment and individual issues in the environment to the economic transformation and development and social governance. In terms of action, a multilevel and multi-subject joint action pattern has taken shape. These practice paths are closely related to different stages of social and economic development and reflect the phased features of sustainable development.

From the practice of urban construction in China, competent ministries and commissions have proposed definitions, established standards and been responsible for assessing the establishment of different types of cities (as shown in Table 8-1). For example, the Ministry of Environmental Protection is responsible for standard setting, trials and assessment for eco-cities and ecological civilization cities; the National Development and Reform Commission is responsible for standard setting, pilot projects and assessment for low-carbon cities and circular economy cities; the Ministry of Housing and Urban-Rural Development is responsible for standard setting, pilot projects and assessment for sponge cities, park cities and eco-park cities; the Ministry of Science and Technology is responsible for standard setting, pilot projects and assessment for innovation demonstration zones on sustainable development and sustainable development experimental zones; the Ministry of Water Resources is responsible for standard setting, pilot projects and assessment for water ecological civilization cities. With clear goals and responsible ministries and commissions, the above pilot and demonstration cities have made good achievements.

4.1. Features and experience of sustainable urban development in China

4.1.1. Shift of focus from individual issues to overall process in development

In 1986, China started to establish "comprehensive social development experimental zones," (which were renamed as sustainable development experimental zones in 1997). In the early stage, the project made a slow progress as evidenced by only twenty-five development experimental zones established from 1986 to 1996.[43] Gradually, the special plans and pilot and demonstration city establishment work led by different ministries and commissions, such as sanitary cities, healthy cities, park cities, eco-cities, environmental protection cities and forest cities, were carried out before and after the UN Conference on Environment and Development one after another. In particular, after the release of *China's Agenda*

Table 8-1 Proposal time and practice of different types of pilot and demonstration cities

Type of city	Time of proposal	Main implementation agency	Practice of pilot project
Comprehensive social development experimental zone was later renamed sustainable development experimental zone	1986	The Ministry of Science and Technology	By the end of 2016, the Ministry of Science and Technology had built and promoted 189 sustainable development experimental zones, covering more than 90% of provinces, municipalities and autonomous regions in China.
Sanitary cities	1989	The Patriotic Health Campaign Committee	By June 23, 2017, 297 cities had been designated as national Sanitary cities in China.
Park cities	1992	The Ministry of Housing and Urban-Rural Development	By 2016, about half of cities had been awarded the title of national Park cities, and 212 counties and forty-seven towns had been awarded the title of national garden counties and towns in China.
Healthy cities	1993	The Patriotic Health Campaign Committee	At the end of 2007, PHCCO designated 10 cities (districts and towns) as the first batch of health pilot cities in China. On November 7, 2016, PHCCO identified thirty-eight national Sanitary cities (districts) as the first batch of national healthy city construction pilot cities.
Eco-cities	1996	The Ministry of Environmental Protection	As of 2015, there were 284 eco-cities in China.
Environmental protection model cities	1997	The Ministry of Environmental Protection	As of June 30, 2017, eighty-four cities (districts) had been awarded the title of national environmental protection model cities, and 129 cities (districts) were being established, covering 30 provinces.

(*continued*)

Table 8-1 Continued

Type of city	Time of proposal	Main implementation agency	Practice of pilot project
Eco-zones	2002	The Ministry of Environmental Protection	As of October 2016, the Ministry of Environmental Protection had awarded more than 110 areas the title of "national ecological cities."
Ecological park cities	2004	The Ministry of Housing and Urban-Rural Development	In 2018, eleven cities became the first batch of national ecological park cities.
Forest cities	2004	The Forestry Administration	As of October 10, 2017, a total of 137 cities across China had been awarded the title of "national forest cities."
Livable cities	2005	The Ministry of Housing and Urban-Rural Development	—
National experimental zones for integrated, complete reform	2005	The National Development and Reform Commission	By June 2013, 12 national experimental zones for integrated, complete reform had been established.
Low-carbon cities	2007	The National Development and Reform Commission	There were a total of 87 pilot cities in three batches, with at least one low-carbon pilot city in 31 provinces in the Chinese mainland.
Ecological civilization cities	2008	The Ministry of Environmental Protection	On September 18, 2017, the Ministry of Environmental Protection awarded 46 cities and counties, including Yanqing District of Beijing, the title of the first batch of national ecological conservation demonstration cities or counties.
Smart cities	2009	The Ministry of Housing and Urban-Rural Development	There were a total of 286 pilot cities in three batches.

Table 8-1 Continued

Type of city	Time of proposal	Main implementation agency	Practice of pilot project
Sponge cities	2012	The Ministry of Housing and Urban-Rural Development	There were a total of 30 pilot cities in two batches.
Water ecological civilization cities	2013	The Ministry of Water Resources	There were a total of 105 pilot cities in two batches.
Circular economy cities	2013	Department of Resource Conservation and Environmental Protection of the National Development and Reform Commission	There were a total of 101 pilot cities in two batches.
Resilient cities	2015	the Ministry of Housing and Urban-Rural Development, the National Development and Reform Commission	There were two types of 58 pilot cities in total.
Trials of new urbanization	2015	the National Development and Reform Commission	From 2015 to the end of 2016, the National Development and Reform Commission approved to establish 248 pilot cities in total.
Innovation demonstration zones on sustainable development	2016	The Ministry of Science and Technology	In April 2018, the State Council officially approved Taiyuan in Shanxi Province, Guilin in Guangxi Zhuang Autonomous Region and Shenzhen in Guangdong Province as the first batch of demonstration zones.

(continued)

Table 8-1 Continued

Type of city	Time of proposal	Main implementation agency	Practice of pilot project
Demonstration areas for the industrial transformation and upgrading	2017	The National Development and Reform Commission, the Ministry of Science and Technology and the Ministry of Industry and Information Technology	In April 2017, the government identified 12 cities (economic zones) as the first batch of demonstration areas for industrial transformation and upgrading.

Source: attributed to the authors.

21, relevant ministries and commissions launched and promoted the establishment of special pilot and demonstration cities relating to sustainable development according to their functions, setting off a new wave of sustainable urban construction aiming at different issues. In the twenty-first century, along with China's accession to the WTO, urbanization and industrialization have been developing rapidly, with explosive emergence of problems in urban environment and urban governance. The root causes for these urban problems are difficult to explain from one side, and the governance couldn't be solved by a single sector inevitably. As a result, some comprehensive pilot projects have been implemented. After "comprehensive social development experimental zones" were renamed as "sustainable development experimental zones," the number of experimental zones increased from 25 to 189 in 2015. In 2015, the "national experimental zones for integrated, complete reform" were launched to meet the internal requirements for complying with the economic globalization and regional economic integration and improving the socialist market economic system, with institutional innovation as the main driving force and the all-round social and economic reform as the main features; the State Council approved twelve national experimental zones for integrated, complete reform. In 2007, low-carbon city project was launched to control carbon emissions, also involving the reform and development of economic structure, energy structure, consumption pattern adjustment and other economic fields; there were eighty-seven pilot provinces and cities in three batches. In 2014, the *National New Urbanization Plan* (2014–2020)[44] (hereinafter referred to as the Plan) proposed a people-oriented urban planning pattern, which was based on efficiency, inclusiveness and sustainability.[45] The Plan explicitly stated that all cities should achieve ecologically sustainable development, while significantly increasing public services. In February 2016, the Urban Climate Change Adaptation Action Plan (hereinafter referred to as the Action Plan),[46] as a part of the *National Climate Change Resilience Strategy*,[47] was officially published to define the concrete actions to achieve sustainable development and promote ecological conservation.[48] In March 2016, the Thirteenth Five-Year Plan clearly required to implement the vision of innovative, coordinated, green, open, and shared development. The Thirteenth Five-Year Plan set out a number of binding goals, covering strengthening resource protection and management, ecological protection and restoration, green financing and green industries, and promoting new-type urbanization, coordinated urban and rural development and coordinated regional development.[49] In September 2016, *China's National Plan on Implementation of the 2030 Agenda for Sustainable Development* was released, which kept the implementation of sustainable development goals line with China's own development emphasis and setting the goal of basically eliminating absolute poverty by 2020.[50] In general, because of the causes for urban problems have become more and more complicated, the ideas and measures to cope with these problems will certainly tend to compound, and comprehensive governance, interdepartmental collaboration and synergy of actions will become the main features of sustainable urban construction, and sustainable urban construction will also be an overall process under the guidance of a top-level design.

4.1.2. Shift of focus from environmental, economic and other survival issues to social issues in development

With rapid advancement of industrialization and urbanization, material wealth of the whole society has been accumulated rapidly. The focus of practice of building sustainable cities in China has gradually evolved from environment, economy and other survival issues, such as healthy cities, Sanitary cities, forest cities, eco-cities, garden city and environmental protection model cities, to development pattern, social equity and social governance pattern. For example, the twelve "national experimental zones for integrated, complete reform" were built to explore a new regional development pattern under new historical conditions based on the overall layout of China's regional economic development; "demonstration areas for the industrial transformation and upgrading for old industrial cities and resource-dependent cities" were established to support transformation and upgrading of traditional robust industries and extend industrial chains, accelerate innovation and cultivation of new techniques and industries, strengthen industrial cooperation, accelerate construction of characteristic industrial parks and industrial clusters, promote the integration of military and civilian sectors, and explore integrated development between industrialization and information technology and between manufacturing and service industries. In the coming ten years, an endogenous driving force mechanism and social support system supporting industrial transformation and upgrading will be established and improved to achieve green transformation and development. Based on inclusiveness and sustainability, the *National New Urbanization Plan* (2014–2020) focused on equal access to public services, farmer-worker's acquiring citizenship and local urbanization, with the aim to improve the accessibility of public services and the fairness of citizens' rights effectively.

4.1.3. Joint action by various actors at different levels

During the development of sustainable cities, China has integrated urban resources through the efforts of the government, non-governmental organizations, enterprises and citizens, so as to create a people-oriented living and working environment and improve the level of sustainable urban development. At the macro level, the central government has played a key guiding role in making strategic plans, issuing preferential industrial policies and building technological innovation systems for sustainable urban development. For example, the sustainable city pilot zones launched in 1986, the *Several Opinions on Promoting Sustainable Development of Resources-oriented Cities* issued by the State Council in 2007 and the Innovative Demonstration Zones for Implementing the 2030 Agenda for Sustainable Development launched in 2016 all pushed and guided the advance of sustainable urban development. At the ministerial level, the National Development and Reform Commission, the Ministry of Housing and Urban-Rural Development, the Ministry of Ecological and Environmental Protection and other ministries and commissions published specific normative documents relating to the construction of sustainable cities, including the National Standards for Low-carbon Cities, the National Standards for Sponge Cities, the National Standards for Livable Cities and the National Standards for Eco-Cities. Some cities in China have also explored and proposed good practices and mechanisms for sustainable urban construction. For instance, in 2010, Shanghai World Expo was successfully held in Shanghai. The Executive Committee of Shanghai World Expo, in cooperation with the UN and the International Exhibitions Bureau, compiled the *Shanghai Manual: A Guide for Sustainable Urban Development in the Twenty-First Century*, providing reference cases and policy guidance for the sustainable development of cities in developing countries.

International organizations have also played a positive role in promoting sustainable urban construction in China. In 2014, Global Environment Facility launched the Sustainable Cities Integrated Approach Pilot Program, in which seven cities in China were selected as global pilot cities. Furthermore, China has actively applied loans from the World Bank, Asian Development Bank, European Investment Bank and International Fund for Agricultural Development and other financial organizations in building a large number of demonstration projects in agriculture, forestry, water and soil conservation, energy, environment, urban construction, and disaster prevention and mitigation, giving a strong boost to sustainable urban development. In July 2015, the UN Environment Programme issued the Guidelines for Evaluation Standard of Sustainable Cities and Communities, and promoted the construction of pilot areas in China.

The public has also played an active role in building sustainable cities. In 2001, China launched environmental warning and education activities and a series of large-scale news interview activities, such as the "Ecological Journey in West China in the New Century" and the "Environmental Protection Journey for South-to-North Water Diversion Project," which objectively reflected the severe environmental and ecological situations and problems and enhanced the public's sense of responsibility and sense of urgency in environmental protection. Meanwhile, it also organized activities to establish "green schools," "green communities" and "green families," gathering more than 40,000 schools.[51] Over the years, workers of Chinese public institutions, teachers and students of universities, environmental organizations and non-governmental organizations, communities and families, women and youth have taken an active part in sustainable development activities through various pilot and demonstration projects and plans. From the environmental publicity, endangered species protection, forest and farmland protection, to the active social supervision, protection of environmental rights and interests, positive advice on green development and many other fields for promoting the sustainable development process, the number of the public actively participating in sustainable urban construction has continued to grow.

4.2. Future development prospects of sustainable cities in China

Over the four decades of reform and opening-up, China has made achievements in sustainable urban construction that are adapted to the level of social and economic development, with abundant experience in construction. In the future, efforts could be made to highlight and emphasize the top-level design role of sustainable urban construction planning in urban development, in order to guide and coordinate urban development. As a developing country, China could also provide its experience in urbanization process, and sustainable urban development and construction to other developing countries for reference.

4.2.1. Top-level design of sustainable urban development is jointly formed dispersion to integration

As sustainable urban development planning involves a wide range of fields and competent departments, it's difficult to coordinate among departments and promote the progress. However, against the backdrop of increasing emphasis on multiple compliance in urban development planning, and integrated and coordinated development, it's a top priority to form the top-level design for urban sustainable development based on synergy from dispersion to integration, which requires joint efforts at multiple levels, mutual coordination between departments and active participation of various parties. Sustainable development planning should be significantly different from special plans led by functional ministries and commissions, and reflect its comprehensiveness and coordination, rather than just any one of multiple demonstration cities or urban development slogans, like "ecological cities," "low-carbon cities," "sponge cities," "park cities" and "circular economy cities." Although these different types of demonstration cities all carry or embody the concept of sustainable development, sustainable cities should become the carrier of all these advanced development concepts and development paths as well as the top-level design for coordinated development of economy, society, environment and spatial layout through multiple compliance. Therefore, sustainable urban development planning should be comprehensive, balanced, dynamic and operable, and need to be formulated and promoted jointly by multiple departments and multiple subjects.

4.2.2. From a follower to a leader, China vigorously promotes ecological conservation and sustainable urban construction

Since the reform and opening-up, China has always been an active participant and contributor in the process of global sustainable development. However, restricted by the economic and social development level, China has mainly followed and participated in the global process, fully learned and gained experience from advanced countries in sustainable urban development, and actively carried out practices. Under the promotion of different ministries and commissions from different perspectives, China has made certain achievements in sustainable urban construction, gradually forming a good momentum of an improved living environment, improved economic structure, better development pattern and more equitable society. The achievements in sustainable urban construction are as follows. Due to rapid urban economic growth, the national economy has maintained sustained and rapid development. A significant progress has been made in urban economic restructuring, as evidenced by an increasing share of the added value of the service sector in GDP, growing citizen consumption rate, narrowed gaps between urban and rural areas and the rate of permanent urban residents up to 58.52%. The urban infrastructure level was improved in an all-round way, and the public service system was basically established and continued to expand its coverage, which indicates further improvements in living standards and quality of urban residents. Significant progress has been made in improving the living environment in cities, including continued reductions in PM2.5, sulfur dioxide, nitrogen oxides, carbon monoxide and other emissions, obvious improvement in energy conservation and environmental protection, an increase in urban green area, and air and water quality. The urban governance pattern has been greatly improved, with a proactive transition from management to governance; a multilevel governance structure with diverse participation has been basically established, so as to constantly improve the equitable rights and interests of urban residents in education, healthcare and development opportunities.

According to the report of the Nineteenth CPC National Congress, under the guidance of the five-pronged overall plan of all-round economic, political, cultural, social and ecological progress and the four-pronged comprehensive strategy of comprehensively building a moderately prosperous society, comprehensively driving reform to a deeper level, comprehensively governing the country in accordance with the law, and comprehensively enforcing strict Party discipline, the CPC will take a two-step approach to build China into a great modern socialist country that is prosperous, strong, democratic, culturally advanced, harmonious and beautiful by the middle of the twenty-first century. As a matter of fact, this has determined the top-level design of sustainable urban construction and development in China. Especially in terms of ecological conservation, the report of the Nineteenth CPC National Congress stated that China should act as an important participant, contributor and leader in global ecological conservation. Undoubtedly, sustainable urban development and ecological conservation are important carriers to implement the spirit of the report of the Nineteenth CPC National Congress; sustainable urban construction will be a specific path to promote green transformation and development and realize ecological civilization.

As man and nature are a living community, sustainable cities should be a place where man and nature coexist in harmony. They should not only create more material and cultural wealth to meet people's ever-growing needs for a better life but also provide more high-quality ecological products to meet people's ever-growing needs for a beautiful environment. This requires cities to transform the development pattern to green development, fully reflect and implement the legal system and policy guidance for accelerating green production and consumption in terms of sustainable urban planning and practice, and establish and improve the green and low-carbon circular economic development system. Moreover, a market-oriented green technology innovation system should be built to develop green finance and strengthen energy conservation, environmental protection, clean production and clean energy industries. Efforts should be made to promote energy production and consumption revolution; build a clean, low-carbon, safe and efficient energy system; improve all-around conservation and recycling of resources, reduce energy and material consumption; achieve the circular linkage between production systems and living system; advocate a simple, moderate, green and low-carbon lifestyle; oppose extravagance, waste and excessive consumption; and call for green travel.

As the largest developing country in the world, China is undergoing urbanization and industrialization. During the forty years of reform and opening-up, the number of cities has increased from 193 to 657 in China, with the urbanization rate rising from 17.92% in 1978 to 58.52% in 2017. China's urbanization process is synchronized with the process of urban construction and expansion, encountering various types of problems faced by almost all countries in the process of urbanization. Therefore, China could share its experience and path of sustainable urban construction and urban ecological civilization construction with other developing countries for reference, and guide and promote the practice of sustainable cities worldwide, becoming a participant, contributor and leader in global green transformation development and global ecological conservation.

5. CONCLUSION

Along with the forty years of reform and opening-up, China's efforts in sustainable urban construction have developed from the construction of comprehensive social development experimental zones in 1986 (which were transformed into sustainable development experimental zones in 1997) to the practice of demonstration areas for the industrial transformation and upgrading in 2017, including a series of urban pilot and demonstration projects, such as sanitary cities, park cities and eco-cities, sponge cities and resilient cities. In general, the process of sustainable city development in China could be divided into three stages. The first stage is from 1986 to 2000, which is the initial practice stage of sustainable urban development; the second stage is from 2001 to 2012, focusing on the transformation and development during sustainable urban construction; the third stage is from 2012 to present, emphasizing coordinated development and paying attention to social equity during sustainable urban construction. According to development features, the focus of sustainable urban practice has gradually shifted from individual topics to overall process, from environment, economy and other survival issues to social issues. Through forty years of development practice, China has formed multilevel, multiagent joint actions and realized multi-field, multi-mechanism coordinated development, with remarkable achievements and development experience in the field of sustainable urban construction. In the future, China will further improve its level of sustainable urban construction from the decentralized planning and development in different sectors to the multi-sectoral, multi-field coordinated development under unified planning. In the future, China will promote ecological conservation, pursue green and low-carbon development, sum up development experience and provide experience for other developing countries, in a bid to contribute to the achievement of global sustainable development goals.

Notes

1. China City Statistical Yearbook 2017.
2. Statistical Communiqué on the 2017 National Economic and Social Development.
3. Ban Ki-moon (2012). Remarks to High Level Delegation of Mayors and Regional Authorities, April 23, 2012.
4. Urban Climate Change Resilience Action Plan, http://www.sdpc.gov.cn/zcfb/zcfbtz/201602/t20160216_774721.html.
5. Pan Jiahua (2012). Strengthening the Establishment of Institutional and Regulatory Regimes of Ecological Civilization, *Finance & Trade Economics*, No. 12.
6. See the UN Sustainable Development Goals, http://www.un.org/sustainabledevelopment/sustainable-development-goal.
7. Announcement of Name List of National Sanitary Cities (Districts) to Be Designated by Patriotic Health Campaign Committee Office during 2012–2014, http://www.nhfpc.gov.cn/jkj/s5899/201501/6af8268344bf4a12a247d1b629a76f9d.shtml.
8. Announcement of Name List of National Sanitary Cities (Districts) to Be Designated by Patriotic Health Campaign Committee Office during 2015–2017, http://www.nhfpc.gov.cn/jkj/s5898/201707/b2e21dfa9d3345249c0d31d5b7656d12.shtml.
9. Proposal on Promoting Healthy Urban Construction in China, http://www.njliaohua.com/lhd_86muq5yvgn3qhty4wk71_1.html.
10. Release of National Evaluation Index System for Healthy Cities, People.cn, http://society.people.com.cn/n1/2018/0410/c1008-29915939.html.
11. Yan Jingsong, Wang Rusong (2004). Connotations, Purpose and Goals of Ecological Cities and Urban Ecological Construction, *Modern Urban Research*, No. 3.
12. Ministry of Housing and Urban-Rural Development of the People's Republic of China, http://www.mohurd.gov.cn/zxydt/201602/t20160201_226501.html, February 1, 2016.
13. Ministry of Housing and Urban-Rural Development of the People's Republic of China, http://www.mohurd.gov.cn/zxydt/201602/t20160201_226501.html, February 1, 2016.
14. State Forestry Administration: China Will Build 6 National-Level Forest City Clusters by 2020, News.163.com, http://news.163.com/18/0127/07/D950722A000187VE.html.
15. Notification on Further Strengthening Exemplary and Leading Role of National Environmental Protection Model Cities in China, China Urban Low-Carbon Economic Network.
16. Jiang Wenjin, Yu Lei, Wang Chengxin et al. (2013). A Study on New Situations of Establishing National Environmental Protection Model Cities, *Environmental Science and Management*, No. 11.
17. Reply to Proposal No. 7794 of the 5th Session of the 12th National People's Congress, http://www.mep.gov.cn/gkml/hbb/jytafw/201709/t20170928_422817.htm.
18. China Announces the First Batch of 12 Industrial Transformation and Upgrading Demonstration Zones, Phoenix News, News.ifeng.com, http://news.ifeng.com/a/20170423/50982889_0.shtml.

19 Five Ministries and Commission, including the National Development and Reform Commission, Issued Notification to Support Construction of the First Batch of Industrial Transformation and Upgrading Demonstration Zones, Gov.cn, http:// www.gov.cn/xinwen/2017-04/21/content_5187958.htm.
20 Zhu Yi (2018). Thoughts on Concept of Smart Urban Construction and Development, *Technology & Market*, No. 3.
21 Xu Jinghua (2012). Comparative Study on Current Situations and Types of Smart Urban Construction in China, *Urban Insight*, No. 4.
22 Guiding Opinions on Promoting Healthy Development of Smart Cities (FGGJ [2014] No. 1770), *Smart City*, September 4, 2014.
23 Yu Kongjian, Li Dihua, Yuan Hong et al. (2015). Theory and Practice of "Sponge Cities", *City Planning Review*, No. 6.
24 Yang Yang, Lin Guangsi (2015). Concept and Thought of Sponge Cities, *South Architecture*, No. 3.
25 China's "Patent Templates" for "Sponge Cities", NetEase News, quoted on December 20, 2015.
26 Interpretation of Construction of "Sponge Cities" in Report on the Work of the Government for Two Sessions, CnR.cn, quoted on March 23, 2017.
27 Liu Hui (2015). Research on Development Pattern and Countermeasures of Urban Circular Economy—A Case Study of Shandong Peninsula Blue Economic Zone, Doctoral dissertation of Liaoning University.
28 Li Caiyun (2016). Research on Spatial-Temporal Evolution of China's Circular Economy Development Level, Doctoral dissertation of Lanzhou University.
29 National Development and Reform Commission Organizes the Establishment of Circular Economy Demonstration Cities (Counties), http://hzs.ndrc.gov.cn/newgzdt/201309/t20130909_557788.html.
30 We Determine 40 Areas as National Circular Economy Demonstration Cities (Counties) in 2013, http://www.ndrc.gov.cn/fzgggz/hjbh/hjjsjyxsh/201312/t20131231_573890.html.
31 Notification on Construction of Circular Economy Demonstration Cities (Counties) (FGHZ [2015] No. 2154), http://hzs.ndrc.gov.cn/newzwxx/201509/t20150924_752108.html.
32 Notification on Designating 61 Areas, Including Tianjin Jinghai County, as National Circular Economy Demonstration Cities (Counties), http://hzs.ndrc.gov.cn/newfzxhjj/xfxd/201601/t20160115_771640.html.
33 Sun Xinzhang (2018). Review and Prospect of Construction of National Sustainable Development Experimental Zones. *China Population, Resources and Environment*, No. 1.
34 State Science and Technology Commission, State Commission for Economic Restructuring: Several Opinions on Establishing Comprehensive Social Development Experimental Zones, 1992.
35 Circular of the State Council on Printing and Issuing the Program for Building the Innovation Demonstration Zones for China's Implementing the 2030 Agenda for Sustainable Development, Government Information Disclosure Column, http://www.gov.cn/zhengce/content/2016-12/13/content_5147412.htm.
36 Weekly Reform News (April 30 to May 6), http://www.ndrc.gov.cn/fzgggz/tzgg/ggkx/201805/ t20180509_885817.html.
37 Pan Jiahua, Ecological Civilization (2015). A New Development Paradigm, *China Economist*, No. 4.

38 Notification on Issuing the Outline of the Thirteenth Five-Year Plan for National Ecological Protection, http://www.scio.gov.cn/xwfbh/xwbfbh/wqfbh/33978/20161212/xgzc35668/Document/1535185/1535185.htm.
39 Wang Jing (2016). Research on Construction Achievements of National Ecological Demonstration Zones, *Heilongjiang Environmental Journal*, No. 4.
40 Hu Qingfang, Wang Yintang, Li Lingjie et al. (2017). A Preliminary Comparison between Water Ecological Civilization Cities and Sponge Cities, *Water Resources Protection*, No. 5.
41 The Construction of the First Batch of 46 Water Ecological Civilization Cities Was Completed, with Sharp Increase in Sewage Disposal Rate, *Urban Roads, Bridges & Flood Control*, No. 2, 2018.
42 Yan Jingsong, Wang Rusong (2004). Connotations, Purpose and Goals of Ecological Cities and Urban Ecological Construction, *Modern Urban Research*, No. 3.
43 Sun Xinzhang (2018). Review and Prospect of Construction of National Sustainable Development Experimental Zones. *China Population, Resources and Environment*, No. 1.
44 National New Urbanization Plan (2014–2020), http://www.gov.cn/zhengce/2014-03/16/content_2640075.htm.
45 World Bank, Development Research Center of the State Council: China: Promoting Efficient, Inclusive and Sustainable Urbanization, https://openknowledge.worldbank.org/handle/10986/18865.
46 Urban Climate Change Resilience Action Plan, see http://www.sdpc.gov.cn/zcfb/zcfbtz/201602/t20160216_774721.html.
47 More information on China's climate commitments, see China's Climate Change Combat of OECD (2016). Available at: http://www.oecd.org/environment/china-climate-change-combat.htm and Annual Report of China's Policies and Actions for Addressing Climate Change of National Development and Reform Commission, http://www.china.com.cn/zhibo/zhuanti/ch-xinwen/2015-11/19/content_37106833.htm.
48 China Business Council for Sustainable Development (2016), see http://english.cbcsd.org.cn/SDtrends/20160325/86084.shtml.
49 The Outline of the Thirteenth Five-Year Plan for National Economic and Social Development of the People's Republic of China, see http://www.gov.cn/xinwen/2016-03/17/content_5054992.htm; http://www.cn.undp.org/content/china/en/home/library/south-south cooperation/13th-five-year-plan--what-to-expect-from-china.html.
50 See http://www.fmprc.gov.cn/web/zyxw/t1405173.shtml.
51 The UN Development Programme: China Sustainable Cities Report 2016: Measuring Ecological Input and Human Development.

CHAPTER NINE

Toward an Ecological Civilization

LI MENG[*]

1. INTRODUCTION

Ecological civilization comes after industrial civilization and above industrial civilization. It is a new form of civilization that follows and absorbs advantages of industrial civilization and represents the sum of material and cultural achievements made by humanity in accordance with the objective law of harmonious development among man, nature and society. Ecological civilization reflects civilization and progress of a society. Ecological conservation aims at a culturally advanced society featuring development of production, an affluent life and a sound ecological environment, with the carrying capacity of resources and the environment as the foundation, the law of nature as the principle and the sustainable development and harmony between man and nature as the goal, rather than abandoning industrial civilization or returning to the primitive production pattern and lifestyle. Ecological conservation is of strategic significance to contemporary China. It bears on the well-being of the Chinese people and the future

---

[*] Li Meng, Doctor, Director and Associate Research Fellow of Research Office for Environmental Economics and Management, Institute of Urban Development and Environment, Chinese Academy of Social Sciences. Her research fields include environmental economy, ecological civilization and sustainable development. She has published five books and more than seventy papers.

of the nation, and is fundamental to the sustainable development of the Chinese nation.

2. DEVELOPMENT HISTORY OF ECOLOGICAL CIVILIZATION

Since the reform and opening-up, the Party has made an active response to the new requirements of the times based on national conditions and international environment during the process of socialist modernization with Chinese characteristics, inherited and developed the Marxist ecological civilization theory, critically absorbed ecological wisdom in traditional Chinese culture, and conducted further explorations and practices in ecological conservation, with the aim to lead the Chinese nation toward green development and great rejuvenation. Considering the status of environmental protection and renewal of development concept, China's ecological conservation could be roughly divided into three stages. In the first stage, environmental protection became a focus in economic development (1978–2002); in the second stage, ecological civilization was built in the course of all-round social progress (2003–2012); in the third stage, ecological conservation was deepened in the context of comprehensive deepening of the reform (2013–present). The three stages are not separated, but related one after the other, which promote China's ecological conservation to become more mature and improved.

2.1. Environmental protection becomes a focus in economic development (1978–2002)

In 1972, the UN held the first global environmental conference in Stockholm, Sweden, and then environmental protection gradually attracted attention from governments around the world. After the reform and opening-up, China has shifted the focus of work to economic construction, while attaching great importance to environmental protection. In view of the environmental disruption caused by extensive economic growth in the early stage of the reform and opening-up, Comrade Deng Xiaoping clearly proposed to attach great importance to environmental protection and governance, repeatedly emphasized that China should pay attention to the rational use of resources and protection of natural environment, and called on the people to "plant trees to green the motherland and benefit future generations." Subsequently, the CPC Central Committee and the State Council made a series of strategic arrangements for environmental protection and environmental renovation, and actively participated in international environmental protection conferences. China took the lead in formulating *China's Agenda 21*, joined multiple world environmental protection organizations, took part in many international environmental conventions, and developed bilateral scientific and technological cooperation and exchanges with Japan, Germany, Korea and other countries, so as to promote sustainable development.

Meanwhile, China began to explore environmental protection by law. The *Constitution of the People's Republic of China*, which was adopted at the 1st Session of the 5th National People's Congress in 1978, put forward environment and natural resources protection and pollution prevention and control for the first time. That is to say, environmental protection was officially written into the Constitution, which laid a foundation for legal construction of environmental protection in China. In 1989, the Eleventh Session of the seventh National People's Congress formally adopted the *Environmental Protection Law of the People's Republic of China*, which is China's first basic law on environmental protection and makes detailed and comprehensive provisions on environmental protection in China. On this basis, China successively formulated and promulgated a number of substantive laws on environmental protection, such as *Forest Law*, *Grassland Law*, *Water Law*, and *Air Pollution Prevention and Control Law*. Currently, China has more than 100 laws and regulations on environmental protection, initially forming a basic framework of legal system on environmental protection in China.

In general, during this period, China's awareness of ecological environment protection gradually grew stronger. The central government increasingly attached great importance to ecological environment protection and construction, and the concept of sustainable development gradually took shape. China built the Three-North Shelterbelt Forest Program, made a number of explorations on pollution prevention, increasingly improved the legal system for ecological environment and strengthened international cooperation on ecological environment, pioneering a path of environmental protection with Chinese characteristics.

2.2. Ecological civilization is built in the course of all-round social progress (2003–2012)

In November 2002, the Sixteenth CPC National Congress put forward the strategic goal of building a well-off society in an all-round way, and stressed to put sustainable development in a prominent position, adhere to the basic national policy of protecting resources and environment, constantly strengthen the sustainable development capacity, improve the ecological environment, significantly enhance the resource utilization efficiency and promote harmony between man and nature, in a bid to push the whole society on a path of civilized development featuring increased production, highly living standards and healthy ecosystem. In October 2003, the third Plenary Session of the 16th CPC Central Committee further defined the *Scientific Outlook on Development*, called for the "Five balances" (balance urban and rural development, development among regions, economic and social development, relations between humankind and nature, and domestic development and openness to the world), and promoted the reform and development of all undertakings. The proposal of sustainable development strategy and scientific outlook on development as well as the implementation of relevant environmental protection principles and concrete measures were milestones in ecological conservation in China.

In 2007, the concept of "ecological civilization" was proposed in the report of the Seventeenth CPC National Congress for the first time. It was another new idea put forward by the Party subsequent to material, cultural-ethical, and political advancement as well as another major theoretical innovation of socialism with Chinese characteristics and the Scientific Outlook on Development of the CPC. Meanwhile, the Seventeenth CPC National Congress put forward the important proposition of building "ecological civilization," and included ecological conservation in the goal of building a well-off society in an all-around way, which required basically forming an energy- and resource-efficient and environment-friendly structure of industries, pattern of growth and mode of consumption. We will have a large-scale circular economy and considerably increase the proportion of renewable energy sources in total energy consumption. The discharge of major pollutants will be brought under effective control, and the ecological and environmental quality will improve notably. In October 2010, the Twelfth Five-Year Plan adopted at the 5th Plenary Session of the 17th Central Committee of the CPC further proposed to establish green, low-carbon development concept, perfect encouragement and restraining mechanisms with an emphasis on energy conservation and on emissions reduction, accelerate the building of resources-saving, environment-friendly production methods and consumption patterns, and advance ecological progress.

From the establishment of the *Scientific Outlook on Development* to the explicit proposal of ecological conservation and its goals, China's ecological environment protection was further deepened, the public's awareness of environmental protection was significantly improved, the concrete work of ecological conservation was arranged level by level, and ecological conservation pilot and demonstration projects were widely carried out at different levels.

2.3. Ecological conservation is deepened in the context of comprehensive deepening of the reform (from 2013 to present)

In November 2012, the Eighteenth CPC National Congress made the strategic decision of "vigorously promoting ecological civilization" at a new historical starting point, raised ecological conservation to an unprecedented strategic height, included it in the overall plan for advancing socialism with Chinese characteristics and formed a five-sphere integrated plan together with economic, political, cultural and social progress, so as to reflect the Party's deeper understanding and more accurate grasp of the implications of socialism with Chinese characteristics, the laws of socialist construction and the laws concerning development of human society. In 2013, the third Plenary Session of the Eighteenth Central Committee of the CPC made further arrangements for ecological conservation, and clearly stated to focus on deepening ecological environment management reform by centering on building a beautiful China. We should accelerate system building to promote ecological progress, improve institutions and mechanisms for developing geographical space, conserve resources and protect the ecological environment, and promote modernization featuring harmonious development between Man and Nature. Later on, the CPC Central Committee and the State Council made systematic and comprehensive arrangements for China's ecological conservation and the ecological progress mechanism and system reform with a series of important documents and guiding plans.

In 2017, the Nineteenth CPC National Congress further emphasized that building an ecological civilization is vital to sustain the Chinese nation's development. China should accelerate the ecological progress system reform, pursue a model of sustainable development featuring increased production, higher living standards and healthy ecosystems. China must continue the *Beautiful China* initiative to create good working and living environments for our people and play its part in ensuring global ecological security. In March 2018, the 1st Session of the Thirteenth National People's Congress adopted the *Amendment to the Constitution of the People's Republic of China* by voting and officially incorporated ecological civilization into the Constitution. This marked that China explored a development path, a rule-of-law model and an environmental protection strategy with Chinese characteristics in line with its national conditions during ecological conservation, and ushered in a new stage of ecological conservation in the new era.

Undoubtedly, China's ecological conservation progressed rapidly in an all-round way in this period. In the course of exploring the path of socialism with Chinese characteristics, China enhanced its understanding of ecological civilization to a new height. As the socialist thought on ecological civilization with Chinese characteristics took shape, China incorporated ecological civilization into major national strategies and goals and established the overall framework for structural reform, with considerable achievements in terms of ecological civilization. Currently, along with the active promotion of ecological civilization in various places, especially the implementation and comprehensive spread of national ecological civilization pilot zones, ecological civilization will become an important part of Chinese socialist progress and a normal state of social development. This means that China's ecological conservation will gradually enter a mature stage, with constant improvement in the overall quality of ecological environment. Ecological civilization will be deeply rooted in the people's awareness and become a cultural habit as well as an important component of contemporary Chinese civilization.

3. ACHIEVEMENTS IN ECOLOGICAL CONSERVATION

3.1. Establishment of the theoretical system ecological conservation

During ecological conservation, China has made constant explorations and accumulated experience, gradually forming and improving its theoretical system. In particular, since the Eighteenth CPC National Congress, the CPC Central Committee with Comrade Xi Jinping at its core has made a series of important statements on ecological conservation in response to the new requirements of the times and practices and the new expectations of the people, formed a systematic and complete ecological conservation theory, gave profound answers to a series of major theoretical and practical questions facing China and the world today in ecological conservation, and established a scientific and complete theoretical system concerning the basic concept of ecological civilization and proposed why and how to build ecological civilization, so as to provide scientific guidelines for ecological conservation.

First, China gives its interpretation of the basic meaning of ecological civilization. Since the Eighteenth CPC National Congress, the CPC Central Committee with Comrade Xi Jinping at its core has put forward a series of new ideas and requirements for ecological conservation. In terms of concept of ecological civilization, it explicitly proposed to respect Nature, follow Nature's laws and protect Nature, adhere to the idea that "green mountains are gold mountains and silver mountains," support value of the environment, environmental capital and spatial balance, and understand that "mountains, rivers, forests, fields, lakes and grasses" are a community of life. In the aspect of the relationship between ecological civilization as well as economic and social development, development and protection should be unified. That is to say, eco-environmental protection and improvement lead to greater productivity, and efforts should be given priority to ecological environment protection. Because a sound ecological environment is the fairest public product and the most inclusive benefits for the people, no welfare is more universally beneficial than a sound natural environment and should work to deliver a greener and better life to the people. In the path of realizing ecological civilization, emphasis should be placed on changing the old thinking and approach of "pollution first, treatment later" and "paying more attention to the end than the source," forming a "green" bottom-line thinking, rule-of-law thinking and system thinking, protecting our natural environment in the same way we would protect our own well-being, and treating ecological environment like treating life, in order to provide people with clean water, fresh air, safe food and a beautiful environment.

Second, the reasons for ecological conservation have been interpreted. Ecological environment is an important part of national or regional comprehensive competitiveness as well as the guarantee and embodiment of people's basic living conditions and quality of life. Environmental protection concerns people's livelihood, development and basic development opportunities, capabilities and interests, and benefits of both current and future generations. In the face of increasingly tight resource constraints, severe environmental pollution and degraded ecosystem, efforts must be made to deepen our awareness and vigorously promote ecological conservation from the strategic perspective of all-round development of socialism with Chinese characteristics and sustainable development of the Chinese nation, in an attempt to open up a new era of socialist ecological progress. In particular, with the progress of society and the improvement of people's living standards, people's demand and requirements for a good ecological environment have also increased, with the rise in the status of ecological environment in people's life happiness index. The core of ecological conservation is to increase the supply of high-quality ecological products, making the sound ecological environment an inclusive benefit for all of the people and a growth area for increasing people's sense of gain and happiness.

Third, the path of ecological conservation has been defined. Chinese President Xi Jinping put forward that "lucid waters and lush mountains are invaluable assets" and proposed the concept of green development, which are fundamental requirements for ecological conservation. They updated traditional understanding of ecology and resources, broke simple ideological restraints that opposite development against protection, defined the methodology for realizing the internal unity of development and protection, mutual promotion and coordinated coexistence, brought profound transformation to development and ideas and ways of governance, and provided the foundational basis for ecological conservation.

Meanwhile, the problem of integration should be solved during ecological conservation. Ecological conservation should be put in a more prominent position; in particular, ecological civilization construction has a new strategic significance in leading four constructions, occupying a "prominent status" in five-sphere integrated plan and forming the "one-in-four" pattern. To this end, ecological conservation should be integrated into economic progress, so as to raise our awareness of the need to promote green, circular and low-carbon development. Ecological conservation should be integrated into political progress, in order to strengthen top-level design and overall planning, and promote institutional innovation of ecological civilization and normalization of legal culture, which is the key and fundamental to integrate ecological civilization into political construction. Ecological conservation should be integrated into cultural and social progress, with the aim to give priority to the most direct and practical problems of the greatest concern to the people. Moreover, it is to promote a new pattern of ecological conservation in which the government plays the leading role, enterprises serve as the main players, the market acts as an effective driving force and the whole society is involved jointly.

In the new era, ecological conservation theory not only provides guidelines for action and fundamental basis for socialist ecological conservation with Chinese characteristics but also offers "Chinese approach" and "oriental wisdom" for the implementation of the *2030 Agenda for Sustainable Development* in the world.

3.2. Breakthroughs in the reform of the system and mechanism of ecological progress

System and mechanism construction is the fundamental guarantee for ecological conservation. In recent years, the central government has issued a series of laws, regulations and institutional documents for promoting environmental governance and ecological conservation, making breakthroughs in the reform of ecological civilization system and mechanism.

First, the top-level design for ecological conservation has been formed. The Eighteenth CPC National Congress incorporated ecological civilization into the five-sphere integrated plan for the socialism with Chinese characteristics. The 3rd Plenary Session of the Eighteenth Central Committee of the CPC proposed to deepen the ecological conservation system reform focusing on building a beautiful China. The fourth Plenary Session of the Eighteenth Central Committee of the CPC required to protect the environment with a strict legal system. The 5th Plenary Session of the Eighteenth Central Committee of the CPC deliberated and adopted the proposal on the Thirteenth Five-Year Plan. In April 2015, the CPC Central Committee and the State Council issued the *Opinions on Accelerating the Ecological Conservation*. In September of the same year, the CPC Central Committee and the State Council issued the *Integrated Reform Plan for Promoting Ecological Progress*. In 2016, the National People's Congress and Chinese People's Political Consultative Conference (NPC and CPPCC) deliberated and approved the *Outline of the Thirteenth Five-Year Plan*. These documents were issued to draw up the top-level design plan of the central government for ecological conservation and chart the course for further advance. Due to limited space, the resolutions on the reform of ecological conservation system and the progress of the top-level design for the reform of China's ecological conservation system at meetings of the Central Leading Group on Comprehensive Reform are shown in Appendix I and Appendix II.

Second, ecological conservation system has been formed rapidly and improved gradually. Among the seventy-nine reform tasks to be completed between 2015 and 2017 defined in the *Integrated Reform Plan for Promoting Ecological Progress*, seventy-three have been completed, and six have been basically completed.[1] At present, efforts have been made to advance the reform for the property rights system of natural resource assets in an orderly way, increasingly strengthen the territorial land development and protection system, launch the spatial planning system reform pilot project in an all-round way, constantly intensify the total resource management and overall conservation system, steadily explore the paid resource use and the system of ecological compensation, strengthen the environmental governance system reform, accelerate the establishment of the environmental governance and ecological protection market system, and basically set up the ecological progress assessment and accountability system. The construction of eight systems in China's ecological civilization system reform is shown in Appendix III.

Third, the legal system of environmental protection has been improved, with unprecedentedly strict supervision. In recent years, the *Land Management Law*, the *Water Pollution Prevention and Control Law*, the *Air Pollution Prevention and Control Law*, the *Wildlife Protection Law*, the *Forest Law*, the *Mineral Resources Law*, and other laws and regulations have been revised and improved; the *Soil Pollution Prevention and Control Law*, the *Basic Law on the Sea*, the *Nuclear Safety Law*, the *Law on Exploration and Development of Resources in Deep Seabed Areas*, and other laws and regulations have been enacted. On January 1, 2015, the new *Environmental Protection Law*, which is "the strictest in history," was implemented. Meanwhile, the law enforcement and supervision of ecological and environmental protection have been stricter than ever before. In order to better implement the concept of ecological civilization effectively, China has started to establish the working mechanism for environmental supervision, strengthen the requirements that both Party and government officials take responsibility for environmental protection, and they both fulfill official duties and uphold clean governance, a supervisory system for auditing the natural resource assets when a relevant official leaves office, and established such systems as the balance sheets for natural resource assets, the audit for outgoing officials' management of natural resource assets, the assessment and examination for regional ecological conservation goals, the lifelong liability accounting system for environmental protection and the accountability for ecological environmental damage for Party and government officials. On the whole, the laws and regulations on environmental protection have been improved, and a rule-of-law atmosphere in strict accordance of laws is taking shape.

Fourth, fruitful achievements have been made in local practice of the ecological conservation system reform in different areas. The first aspect is the red line system for ecological protection. (1) By specific category and zoning, each province set up the red line protection zones in strict accordance with their actual conditions. Hubei Province divided ecological red line zones into four categories, namely water conservation zone, biodiversity conservation zone, soil conservation zone, and flood control and storage zone for lakes and wetlands in the middle reaches of the Yangtze River. Chongqing's ecological red line zones included "four-mountain" forbidden construction zones, Three Gorges Reservoir fluctuation zone and non-commercial ecological forest with Chongqing's local characteristics. (2) In terms of management measures, most provinces implemented differentiated management for red line zones based on importance degree and grading. Shanghai, Jiangsu, Hubei, Hunan, Shaanxi and many other provinces divided the ecological red line zones into first grade control zones and second grade control zones. (3) In the aspect of punishment and accountability mechanisms, Sichuan Province carries out performance assessment on the effectiveness of red line protection zones regularly, and took the assessment results as the direct basis for determining ecological compensation funds. Guizhou Province established the ecological protection red line accountability system for leading officials during their term of office, and those who fail to protect the environment or make wrong decisions that lead to ecological damage would be held accountable.

The second aspect is the resource and environmental tax reform. In May 2016, the Ministry of Finance and the State Taxation Administration issued the *Notification on Comprehensively Promoting Resource Tax Reform* and the *Interim Measures for Water Resource Tax Reform Pilot Projects*, and specified to promote the resource tax reform in an all-round way since July 1, 2016. The *Environmental Protection Tax Law* took effect on January 1, 2018. At the practical level of the resource and environment tax reform, Hebei Province carried out the water resource tax pilot projects, and included surface water and groundwater in the scope of taxation by the water resource fee-to-tax method. The tax rates for industries with a high water consumption, over limit water consumption and groundwater intaking in groundwater overexploitation zones should be increased appropriately.

The third aspect is biodiversity conservation. Local governments also made active explorations and improvements in mechanisms and systems. For example, Yunnan Province took all-level linkage measures, strengthened biodiversity protection in an all-round way, established a biodiversity research institute and a biodiversity protection fund, continued to implement the *Yunnan Strategy and Action Plan on Biodiversity Protection* (2012–2030), took the lead in releasing the provincial-level directory of biological species and red list directory in China, wrote the *Encyclopedia of Yunnan* (by Ecology), compiled and implemented the *Schematic Design for Major National Biodiversity Protection Project in Yunnan Province*, and issued the *Regulations on Biodiversity Protection in Yunnan Province* for legislative protection.

3.3. Constant deepening of practice in ecological environment protection and governance

First, the environmental protection co-governance pattern is taking shape. China's unbalanced economic and social development and complicated environmental problems determine the multi-dimensional choice of the environmental management pattern. In terms of the main architecture, local Party committees, local governments, local people's congresses, local CPPCC committees, judiciary authorities, social organizations, enterprises and individuals have played an orderly role in ecological conservation through the reform, and an environmental co-governance pattern is taking shape. For instance, a list of well-defined government powers was established to define the principles of the integration of power and responsibility and the lifelong accountability. Environmental protection assessment, supervision, inspection, negotiation and accountability were conducted to promote the in-depth implementation of the same duties shared by the party and the government. Environmental protection enterprises, especially leading ones, took an active part in the third-party environmental protection governance through investment and financing mechanisms, so as to improve environmental quality. Citizens and social organizations strengthened supervision over enterprises and law enforcement agencies based on information disclosure, making people feel the environmental improvement effect through concrete and vivid ecological civilization practices.

Second, regional development quality has been improved continuously. In terms of regional development, related resources were coordinated and improved through reform. In order to reduce logistics costs and logistics time and promote industrial structure optimization, adjustment and even integrated development in a larger regional scope, local governments of the Beijing-Tianjin-Hebei region, the Yangtze River Delta and the Pearl River Delta strengthened construction of regional transportation networks; and those of the Beijing-Tianjin-Hebei region formulated the regional coordinated development plan, and adopted unified planning, standards and monitoring, coordinated law enforcement, unified judicial administration and emergency response measures. Hainan Province, Ningxia Hui Autonomous Region and some counties and cities have been carrying out the "multiple compliance" pilot projects, with constant improvement in development space. Ecological compensation mechanisms for areas between upper and lower reaches, and such key zones as forest, grassland, wetland, desert, ocean, river and cultivated land, forbidden development zones and key ecological function zones have been established in an all-round way. The fair mechanism for regional green development has begun to take effect.

Pollution rights trading, carbon emissions trading, water right trading and energy-using right trading have been ongoing. Third-party governance markets of sewage and garbage disposal have continued to boom, laying a market foundation for improving regional environmental quality. Steady progress has been made in urban waste sorting and centralized sewage treatment. Breakthroughs have been made in waste sorting, collection and disposal, and sewage treatment in some provinces. New progress has been made in integrated urban-rural environmental renovation, marking that the civilization awareness and level of the society have entered a new stage. Through the regional integrated green development reform, the quality and efficiency of regional economic development have been improved as a whole, with gradual coincidence between the sociality and naturality of environmental protection.

Third, prominent environmental problems have been alleviated. Ecological conservation has been incorporated into the five-sphere integrated plan for building socialism with Chinese characteristics. The CPC Central Committee with Comrade Xi Jinping at its core has made a top-level design and overall plan for ecological civilization construction and established a national governance system for ecological conservation. In recent years, China has made all-round progress in ecological conservation, with significantly accelerated green development and alleviated environmental problems. Specifically, notable achievements have been made in circular economic development, the ambient air quality and surface water quality have improved, serious ecosystem degradation has been initially curbed, and urban and rural living environment has improved gradually. Due to limited space, the specific data and analysis are shown in Appendix IV.

Through analysis, the achievements in China's ecological conservation are mainly benefited from four aspects. First, the central government has played a powerful guiding and promoting role, conducted problem-oriented and goal-oriented top-level design, and defined the reform direction. For example, the ecological civilization system and mechanism reform targeted at bottlenecks and weak links with precise focus and concerted efforts. Institutional gaps have been filled with the reforms for the property rights system of natural resource assets, the spatial planning system and the environmental governance and ecological restoration market system, and systems and working mechanisms have been innovated constantly, thus forming a good situation of mutual coordination and support. Besides, the supply-side structural reform and the government function reform have been guided and driven by means of assessment, so as to directly advance the economic system reform and the financial reform, and promote the construction of a green, circular and low-carbon industrial system. Second, local governments have made explorations, practices and innovations for typical demonstration, sharing and promotion. As of 2016, forty-two national low-carbon pilot provinces and municipalities were established throughout China, sixteen ecological provinces, autonomous regions and municipalities were set up, more than 1,000 cities, counties and districts were promoting the construction of eco-cities and eco-counties, and ninety-two ecological cities and counties and 4,596 ecological townships and towns with a beautiful environment were designated. In China, seventy-two ecological conservation pilot projects were established in six batches; the first batch of fifty-seven areas were approved to be ecological civilization demonstration zones; local governments of thirteen cities (prefectures and cities), including Ulanqab in the Inner Mongolia Autonomous Region, and seventy-four counties (cities, districts, counties and leagues), including Wushan County in Chongqing, launched the national ecological conservation demonstration and pilot projects; twenty-six national eco-industrial demonstration parks passed the acceptance inspection and were named upon approval. In addition, 76 national ecotourism demonstration zones, 105 national water ecological progress pilot cities and more than 170 marine protected zones were established. In 2017, the national ecological civilization pilot zone was set up to integrate resources and platforms and improve the quality of pilot projects. As a result, local ecological environment tended to get better, with significantly improvement in quality and efficiency of economic development. Third, efforts have been made to improve the rule of law, regulate behaviors, deepen institutional reform and provide strong support. Ecological conservation has always been promoted by deepening the system and mechanism reform incorporating it into law and making institutional arrangements. The new *Environmental Protection Law* has included a strict assessment mechanism, specified such punishment measures as "administrative detention," "resignation due to blame" and "daily punishment," and contained the ecological protection red line for the first time to protect citizens' right to participate in environmental protection. The *Measures on Evaluating Performance in Advancing Ecological Progress* are objective, fair, scientific and standardized, focus on key points and practical results, and define both rewards and punishments, playing a powerful guiding and restraining role in regulating and constraining relevant decisions. The amendment to the *Water Pollution Prevention and Control Law* proposed that EIA units, testing institutions and pollutant discharging units should bear joint liability, which was of great significance to eliminate inadequate EIA and other deep-rooted malpractices. Along with the deepening of the ecological conservation system reform, relevant systems and mechanisms have also been improved. Fourth, supervision, inspection, targeted measures and category-based assessment have been provided responsibilities, rights and interests. During the Twelfth Five-Year Plan period, the government strengthened inspection and supervision over ecological environment protection and other ecological conservation work. In 2016, the Central Supervision and Inspection Group for Environmental Protection composed of relevant leaders of the CPC Central Commission for Discipline Inspection and the Organization Department of the CPC Central Committee under the leadership of the Ministry of Environmental Protection set off rounds of "inspection storms" over sixteen provinces (autonomous regions and municipalities), solved up to 15,761 environmental cases reported by the public, and held more than 4,400 people accountable, acting as a strong deterrent to illegal behaviors that damage or undermine ecological conservation. In the meantime, efforts were made to conduct assessments by category and establish a responsibility framework in which the Party and the government share responsibilities and provide joint management, in a bid to give full play to the leading role of party committees and the synergistic effect between the Party and the government. Recently, a number of measures have been introduced to ensure the sustainability and normalization of supervision and inspections.

4. PROBLEMS AND CHALLENGES FACING ECOLOGICAL CONSERVATION

4.1. Remaining problems

Although China has made many achievements in ecological conservation, there are still some problems. The protruding problems are as follows.

First, basic data was insufficient, distorted and inaccurate. Due to the heavy workload in basic data acquisition, insufficient input and difficulties in supervision and administration, there were frauds on the whole, with data distortion and inaccuracy. Apart from environmental monitoring data frauds, there were also frauds in the implementation of overcapacity cut. This reflected inadequate coordination for relevant work, and there have been loopholes in the systems that need to be corrected and improved urgently. Therefore, efforts were made to strengthen verification and review of data submitted, in order to prevent such incidents.

Second, reform tasks were less coordinated and operable, resulting in inadequate implementation. After the main framework of the ecological conservation system reform was established, the details have still been improved, and it took time to coordinate and support the reform tasks. In terms of time arrangement, many reform tasks were completed during the Thirteenth Five-Year Plan period. However, lagging basic reforms made it difficult to implement some reform measures issued. In addition to theoretical problems, such as the unclear basic concepts like boundary of property rights and the public ownership remaining to be further clarified, there were still other problems, such as incomplete basic measurement data as well as inadequate measurement techniques and methods still in exploration.

Third, there were big demand gaps in capital and talents. In order to achieve China's environmental pollution control goal and the international commitment of peaking its carbon emissions by 2030 or before, it was estimated to need RMB3 trillion to RMB4 trillion of green investment every year, which couldn't be fully covered by financial funds. Accordingly, green finance should be further developed and expanded, and attention should be paid to preventing financial risks. Supervision and efficiency assessment over the process of capital use should be strengthened. Insufficient professionals and team building were also hard to meet business demands.

Fourth, pseudo-ecological conservation was prominent. In recent years, many cities attached great importance to advancing ecological conservation, and all sectors took ecological governance as an important task as well. However, differences in ideas and concerns among various sectors brought distinct governance strategies. Worse still, many cities carried out unrealistic projects pursued solely for show and pseudo-ecological behaviors that destroyed ecosystems. For example, watercourses were fully hardened in the name of flood control; fallen leaves in urban woodlands were cleaned in the name of fire prevention and sanitation; the original natural landscape was transformed in the name of governance and upgrading; the urban greening pattern was transplanted to the countryside in the name of afforestation. At last, the original ecological landscape was lost.

4.2. Challenges

Due to the complexity of the social and economic environment and other factors, there are also challenges in China's ecological conservation at present, in addition to the problems.

First, how to transfer lucid waters and lush mountains into invaluable assets? In some areas, especially economically underdeveloped central and western areas with abundant ecological assets or a fragile ecological environment, people's income and quality of life still lagged far behind those in the eastern area. In both awareness and practice, it was difficult to carry out the scientific conclusion that "lucid waters and lush mountains are invaluable assets." Even in the eastern area, there were still some short-sighted behaviors to sacrifice lucid waters and lush mountains for wealth. The lingering haze on a large scale was caused by the behaviors of sacrificing other's long-term interests for the sake of immediate self-interests to a large extent.

Second, how to put green development into practice? The problems including how to convert ecological advantages into development advantages, connect ecological demands with economic demands, turn the pressure of filling in ecological gaps into the impetus for development and transformation, make ecological economy into a powerful engine for promoting and driving economic development and transformation, increasing both economic development and quality, increasing employment, building a strong country and delivering a better life for peoples, and avoiding falling into the driving force of "middle-income trap" were to be solved in green development. At present, the level of green development was mainly driven by policies. In addition to policy orientation, green development also required financial support, technological support and promotion of social equity. At present, China's green development is still in its infancy. According to objective situations at home and abroad, green development required long-term unremitting efforts.

Third, how to better play the roles of market mechanism? In some reform schemes, market mechanism was confused with the concept of the market, and its roles were simplified into the establishment of the market and the introduction of third parties in environmental management, environmental monitoring and ecological restoration. This blurred the responsible parties and increased the risks of forming interest chains and giving rise to corruption, rather than reducing the government's administrative duties. How to guarantee the normal operation of market mechanism should be further explored in terms of systems and mechanisms, and specific operation patterns.

Fourth, how to cope with uncertainties in ecological environmental governance? Take haze control as an example. Due to the complex formation mechanism, the lower emission reduction elasticity, the unchanged main meteorological conditions and the uncertainty about whether static and stable meteorological conditions significantly reduced, there was still uncertainty about the governance effect in the short term, which couldn't be solved by a commitment. Efforts should be made to implement stricter air pollutant concentration and total emission control standards and expand the coverage of pollution source control, so as to effectively curb the trend of air quality deterioration. It is necessary to strengthen basic scientific research, and intensify and refine management, which is a heavy workload and very difficult.

Fifth, how to diversify ecological conservation patterns? With diverse ecological conservation patterns, every place has the opportunity to build ecological civilization with its own regional characteristics. Cultural advancement reflects distinct national and regional characteristics. As a kind of civilization, ecological civilization should not follow unified standards as well, and there shouldn't be a pattern for ecological conservation. How to promote the integration of ecological conservation with local culture and reflect local characteristics of ecological civilization construction was also one of the questions to be considered.

5. POLICY SUGGESTIONS FOR DEEPENING REFORM OF ECOLOGICAL CONSERVATION SYSTEM

5.1. Key tasks in ecological conservation in the time to come

In October 2017, the Nineteenth CPC National Congress was held, which was a milestone in the reform of China's ecological conservation system.

The Nineteenth CPC National Congress attached great importance to ecological civilization. Among the thirteen parts of the report to the Nineteenth CPC National Congress, one part was dedicated ecological civilization. In the whole text of the report, "ecological civilization" was directly mentioned twelve times, "ecology" was mentioned forty-five times, "green" was mentioned fifteen times and "beautiful" was mentioned eight times. The report put forward a series of new ideas, new requirements, new goals and new arrangements for ecological conservation. According to the report to the Nineteenth CPC National Congress, key tasks of ecological conservation in the time to come included promoting green development, solving prominent environmental problems, intensifying the protection of ecosystems and reforming the environmental regulation system. Among the four tasks, the second task focused on environmental protection; the third task focused on ecological protection; the first task integrated the above two tasks and summarized the overall thinking of the development of the whole economic life from a comprehensive perspective of industrial development; and the fourth task proposed an effective control system of the central government on environmental protection and ecosystem, which was one of the key supporting points that really put the first three tasks into practice. Besides, the report to the Nineteenth CPC National Congress proposed strategic goals and positioning of "winning the battle against air pollution" and becoming "an important participant, contributor, and torchbearer in the global endeavor for ecological civilization," which actually put forward higher requirements for the ecological civilization system reform. Appendix V illustrates the new arrangements and key tasks in ecological conservation in the time to come set forth in the report to the Nineteenth CPC National Congress.

In general, in the time to come, the key to further promoting China's ecological conservation lies in further deepening China's ecological civilization system reform, which focuses on strengthening top-level design, intensifying long-term mechanisms and establishing market mechanisms.

First, a top-level design should be further improved. Although remarkable achievements have been made in China's ecological conservation, some prominent ecological environmental problems have remained to be well solved, including winning the battle against air pollution, environmental restoration and protection in the Yangtze River Economic Belt, and increasingly prominent rural pollution control.

Rural pollution control could be taken as an example. At present, China's environmental protection work has focused more on industrial and urban pollution than rural pollution prevention and control. In recent years, the central government has increased investment in addressing sewage and garbage problems in rural areas, which, however, is far from implementing rural pollution prevention and control. As for the weak link of rural pollution prevention and control, the environmental protection department and relevant departments should first issue a comprehensive top-level design and arrangements as soon as possible in the future, taking rural environmental pollution prevention and control as an important component of the overall rural revitalization plan.

Second, long-term mechanisms should be established and improved. The central government inspections on environmental protection were a major reform measure of ecological civilization systems and mechanisms promoted by Chinese President Xi Jinping personally after the Eighteenth CPC National Congress. The supervision and inspection pilot project started in Hebei Province at the end of 2015, and covered all of the thirty-one provinces (autonomous regions and municipalities) before the Nineteenth CPC National Congress. The central government inspections on environmental protection have effectively promoted the establishment and improvement of local environmental protection and ecological civilization mechanisms, and forced local governments to accelerate the establishment and improvement of relevant laws and regulations. Although the central environmental protection supervision and inspection made such significant achievements, how to establish and maintain normal and long-term supervision and inspection mechanisms on environmental protection based on the existing practice is an important guarantee to maintain the existing achievements of ecological conservation as well as a key point required by the ecological civilization system reform in the next stage.

Meanwhile, it's urgent to establish long-term environmental control mechanisms at present, bring the guiding role of environmental control into play in green development, effectively guide enterprise transformation and upgrading, promote technical innovation, and move toward green production, while encouraging development of green industries, strengthening energy-saving and environmental protection industries, clean production industries and clean energy industries, and making green industries into substitute industries in economic growth. Environmental control work should cover different enterprises, different industries and different area types, and adopt different environmental control measures, so as to give consideration to both economic growth and enterprise competitiveness, achieve both goals of environmental protection and economic development, and promote green development in a real sense.

Third, market mechanisms should be rationally used. Over the years, China's measures in ecological conservation have mainly existed in strengthening government supervision in an all-round way. The connotation of ecological conservation is to realize coordinated development of economy and environment protection. Nevertheless, win-win development of economy and environment could not be achieved only by relying on the government's promotion and supervision. It's necessary to establish improved market mechanisms and incentive mechanisms. However, China's efforts in this aspect remain to be strengthened at present, which is also a key point in ecological conservation and system reform in the future.

5.2. Policy suggestions on deepening ecological conservation reform

In the time to come, the key to further promote ecological conservation in China is to deepen the ecological civilization mechanism and system reform, which is also an important guarantee for achieving green development, solving prominent environmental problems and strengthening ecosystem protection. Based on problems and key points in ecological conservation at present, it's suggested to give "problem-orientated," "goal-orientated" and "market-orientated" guidance; deepen management system, evaluation mechanism and market mechanism reforms effectively; promote the implementation of the ecological civilization system and mechanism reform; and improve the executive force of the reform and release vitality and dividends of the reform, in a bid to bring China into a new era of ecological civilization.

5.2.1. Break through the blockades of interests, grasp the "problem-oriented" approach, innovate the supervision system on ecological environmental protection and improve the executive force of the reform

At present, the ecological civilization system reform has entered the late period of operation and implementation. However, due to irrational arrangements of the vested environmental protection system, environmental protection agencies have suffered from decentralized functions, mismatched powers and responsibilities, and weak law enforcement and inadequate supervision. Multiple strategies and arrangements of the system reform still stay in documents and clauses.

Regulation refers to both supervision and administration. Therefore, the regulatory body of ecological environment should be composed of three parties: first, the Ministry of Ecology and Environment for supervision and administration as a professional government administration; second, environmental public welfare organizations and other social organizations for supervision; and third, the general public for supervision. In terms of social organizations and public supervision, there was a lack of institutional guarantee, with inadequate supervision and no supervision channel, which resulted in failure in supervision. In order to reform the ecological environment supervision system, it is necessary to implement a "problem-oriented" supervision system reform pattern.

(1) The property rights system of natural resource assets should be clarified to promote the separation of ownership, right to use and regulatory power. Clearly established ownership is an important prerequisite for the

implementation of asset management of natural resources, the foundation for establishing the attribution order of natural resource rights as well as a necessary condition for realizing the transfer order of natural resource rights. Therefore, China should first establish and eventually form a natural resource asset property right system with clear attribution, defined rights and responsibilities, and effective supervision. Efforts should be made to complete the registration of natural resource rights as soon as possible and define the owners and users of natural resources, so as to clarity the responsible party for supervision and administration of natural resources. The tenure of use should be extended appropriately to adapt to the economic cycle of resource exploitation and utilization; frequent adjustments of the right to use should be avoided as far as possible; in particular, long-term stable right to use should be ensured for land, forestry and other long-cycle resources. Currently, with integrated functions of eight departments, the new Ministry of Natural Resources was established to implement the central government's requirements for uniformly performing the duties of the whole people as the owner of natural resource assets, provide scientific guidance for protection and rational development of natural resources, and complete the registration of the authentic right of natural resources, in a bid to ensure paid use and rational development and use of natural resources, and conduct unified supervision, protection and restoration of natural resources.

Responsibilities, rights and interests in resource management among departments of the central government as well as between the central government and local governments should be defined to prevent the loss of natural resource assets caused by lax management of resource property rights or property right disputes. Natural resource assets should be managed on the basis of unified rules and regulations. However, in specific management of natural resource assets, differentiated and classified management should be carried out according to different types of natural resources, their characteristics, regions and industries. Meanwhile, the principle of separation of ownership and management right should be followed in the active improvement of market mechanisms and the establishment of diversified management patterns of natural resource assets.

(2) The environmental protection management system should be improved to define the responsibility boundaries of all departments. In terms of ecological environment protection, efforts should be made to unify environmental decisions and supervision, implement professional work division with separate responsibilities, and define the responsibility boundaries of various agencies and departments. In 2015, the CPC

Central Committee and the State Council issued the *Integrated Reform Plan for Promoting Ecological Progress*, which clearly proposed to "improving the administrative system for environmental protection. An effective administrative system for environmental protection will be established to strictly regulate the emissions of all pollutants. Duties and responsibilities for environmental protection, which are currently spread across departments, will be assigned to one single department, progressively creating a system whereby one department is responsible for unified regulation and administrative law enforcement over urban and rural environmental protection work."

There were too many departments in charge of environmental protection in China, which inevitably led to such problems as decentralized administrative functions and confused authorities in environmental protection. In turn, the chaotic phenomena also weakened the power and functions of environmental protection departments in China. Founded in March 2018, the Ministry of Ecology and Environment is responsible for unifying supervision and administration of ecological and environmental protection. Meanwhile, in the principle of the separation of decision-making power, executive power and supervision power, this department focuses on unifying top-level designs, law enforcement and supervision. Specific administrative execution should be divided by professional organizations of different places, which should be responsible for environmental protection issues in their professional fields, and whose duty scope and boundaries should be defined through legislation.

(3) Relevant laws should be revised and improved, and a comprehensive legal and policy system should be established. At the legal level, the existing legal system should be reviewed and improved according to the requirements of the ecological civilization system reform. The *Land Management Law*, the *Mineral Resources Law*, the *Water Law*, the *Grassland Law*, the *Forest Law* and other resources-related laws and regulations should be revised and improved rapidly to strengthen comprehensive legislation on natural resources. In particular, central and local, domestic and international legal systems, which were based on the *Constitution* and the *Environmental Protection Law* and directly involved resource conservation and environment friendliness, should be revised, supplemented and improved. Besides, law enforcement should be strengthened to create an atmosphere that it's glorious to abide by the laws and shameful to break the laws in public opinions. At the policy level, industrial policy, consumption policy, population policy, land policy, financial policy and fiscal policy should further reflect the orientation of the ecological civilization

system reform. Meanwhile, a series of environmental and economic policies should be established as soon as possible, with ecological compensation policy, emission trading policy, environmental taxation policy, green credit, insurance and securities policy as the main contents.

(4) Public participation mechanisms for natural resources and ecological environment protection should be established and improved. Publicity and education of natural resources knowledge among the public should be strengthened continuously to make them realize the scarcity of natural resources and the necessity of protecting resources and enhance the public's sense of ownership of resources and environment in an all-round way. Based on the relations between the public management of natural resource assets and the vital interests of individual business management, the public's values of sustainable use of resources should be cultivated to improve the public's initiative to consciously participate in the management of natural resources. Social organizations should cultivate and develop their role in protecting ecological environment. In particular, environmental organizations have unique advantages in environmental management. First, they are uncorrelated with interests and relatively fair; second, they are simple in hierarchy, with a high efficiency; third, they are highly professional; fourth, they are good at increasing people's sense of participation and their awareness of environmental protection. Compared with the government and enterprises in the relationship between supervisor and supervisee, environmental protection organizations play a bridge role in regulating the relationship between the government and enterprises. British environmental protection organizations have played an important role in setting standards, supervising the government and enterprises, and organizing technical exchanges and personnel training. In comparison, China's social organizations have a late start in development and aren't professional enough. Therefore, efforts should be made to further give play to the roles of environmental protection organizations and make social organizations become advisers, helpers and supervisors of the government in environmental governance.

5.2.2. Take root in green development, strengthen "goal-oriented" development and deepen the assessment mechanism of ecological conservation to ensure the implementation of the reform

During the ecological civilization system reform, the obstacles in the implementation of policies and measures mainly came from the complicated authority division of administrative departments at all levels as well as different interest appeals. At present, the ecological civilization system reform should be promoted quickly, which shouldn't be impacted by authority division. In the reform, a roundabout strategy should be adopted based on assessment goals, so as to guide and encourage governments at all levels to actively promote ecological conservation.

(1) Comprehensive performance assessment criteria should be established. There are two meanings in the so-called comprehensive performance assessment criteria. First, performance shouldn't be assessed only based on economic conditions, but also based on Gini coefficient, Engel coefficient, urban-rural dual structure index, human development index and environmental index. Second, economic conditions shouldn't be assessed only based on economic growth speed and scale, but also based on economic structure, quality and efficiency, energy consumption, water consumption and material consumption per unit of GDP, pollutant discharge level per unit of GDP, labor productivity and economic capacity per unit of land area. Only in this way can we achieve all of increased production, highly living standards and healthy ecosystem and give both consideration to economic development and strategic goal of "winning the battle against air pollution."

(2) Reward and punishment mechanisms should be based on the ecological conservation assessment results. The assessment results should be adopted in improving work, strengthening the daily inspection and year-end assessment of ecological conservation, finding the problems and urging relevant units to rectify them in a timely manner. Reward and punishment mechanisms should be established to commend and reward units and individuals that made outstanding contributions to ecological conservation with the assessment results as an important basis for promoting and appointing leading cadres. The lifelong accountability system for ecological and environmental damage should be established effective, and must hold to account on a permanent basis the leading officials who cause serious consequences through blind decisions that disregard the environment.

(3) An assessment mechanism with multi-sectoral participation should be established. The ecological conservation system reform is a huge and systematic project involving economy, politics, culture, society and ecology, which requires strengthened organizational leadership and joint management. It is suggested that the central government determines leading departments and responsible departments, and establish an assessment mechanism composed of organizational department, comprehensive economic department, environmental protection department, statistics department, supervision department and other departments, so as to

comprehensively deepen and promote the ecological civilization assessment mechanism reform of ecological civilization.

5.2.3. Adhere to ecological guidance, strengthen the "market-oriented" mechanism, build the ecological industry development mechanism, and release the dividends of the reform

The fundamental purpose of the ecological civilization system reform is to promote the national governance system in the field of ecological civilization, modernize the governance capabilities, solve people's ever-growing needs for a better life and the contradictions in unbalanced and inadequate development, and provide more high-quality ecological products, in an attempt to usher into a new era of socialist ecological civilization.

So far, general framework, like the beams and pillars of a house, have initially taken shape in China's ecological conservation. The spatial planning system reform pilot project was launched in an all-round way to constantly strengthen the total resource management and overall economy system, make steady progress in the paid resource use and ecological compensation system, intensify the environmental governance system reform, and accelerate the establishment of the market governance and ecological protection market system. The ecological civilization performance assessment and accountability system has been set up. However, these policies focused on top-level designs, especially in the aspect of administrative performance assessment. In terms of specific market mechanisms, there are still many fields to be improved and refined, especially in stimulation of market vitality. Therefore, the policy focus in the next stage should gradually shift to improving the market mechanism for ecological conservation, especially in operability and incentive, and relevant mechanisms should be established according to market demands.

As a brand-new market, ecological civilization market generally includes ecological products market, emission trading market, environmental information market, environmental consulting market and environmental protection service market in general. The socialist market economy system should be established based on a unified, open, competitive and orderly market system. In a word, the market system is an organic unity composed of various interconnected markets. Under the new circumstances, we should actively establish a new way of thinking, actively study new laws and strive to find new methods, so as to establish and improve the ecological civilization market system and deepen the reform and innovation of the ecological civilization system.

(1) Accurate policy guidance is required to give full play to the roles of market mechanisms. To give full play to the roles of market mechanisms, we

must learn and be good at solving problems by economic means, and put more emphasis on the precision of policies. For example, the collection of sewage charges in environmental law enforcement has generally been considered to be less rigid. If environmental law enforcement is changed into tax law enforcement, market mechanisms will play a more stable role. Next, we should accelerate the environmental tax reform, promote the improvement of existing fiscal and tax policies, achieve comprehensive coverage at overall and strategic levels, prevent policy "vacuum" in some fields, and correctly grasp the relationship between improving policy systems and highlighting policy priorities. In the meantime, market mechanisms should be activated to promote the release of ecological dividends. Since the reform and opening-up, the driving force of China's economic development has been derived from the "demographic dividend" to a large extent, which has been realized and amplified through urbanization and industrialization. At present, the universal two-child policy failed to reverse the trend of population aging, and the source of demographic dividend has nearly exhausted. The new engines of economic and social development in the era of ecological civilization should be the release and amplification of ecological dividends, which could only come true through industrialization and marketization. This is the focus of our ecological civilization system reform in the next stage.

(2) The roles of market mechanisms in revitalizing natural resource assets and adding value should be brought into play gradually. Natural resources will get involved in distribution and share development dividends through policy guidance. At present, natural resource assets could be restored gradually through natural rehabilitation and optimized ecological restoration, which, however, couldn't fully meet sustainable development demands of positive interactions of ecosystems. As natural resources are an important part of ecosystems, their stock and quality should be maintained and value-added, and their products and income should be shared by man and the nature. Natural resources should get involved in the distribution system of national economy to form an income distribution mechanism for ecological civilization with the harmony between man and the nature, while expanding the market space and potential in ecological conservation.

Natural resource asset management is an inevitable requirement for revitalizing natural resource assets and safeguarding national owner's rights and interests, as well as an effective guarantee for rational development and utilization of natural resources. Natural resource asset management requires the market to play an active role in effective and rational allocation of natural resources, so as to maximize the enthusiasm of main market players. In China, the government sector has always been regarded as the subject of sustainable natural resource management. However, the practice of natural resource management has proved that private individuals or manufacturers are the key to solve resource-related problems. In addition to accelerating the primary task of improving the natural resource property right system, efforts should also be made to gradually establish natural resource asset trading markets in China and key areas based on pilot projects, accelerate the reform of natural resources and their product prices, expand the coverage of resources tax, and gradually improve the paid resource use system so that resource development and utilization could reflect the degree of resource scarcity and the costs of ecological environment damage and restoration.

(3) Market mechanisms should be innovated to give positive incentives to the transformation of green drivers and the construction of long-term mechanisms. For a long time, measures taken in China's ecological conservation have mainly reflected all-round improvement in government regulation. The connotation of ecological conservation is to realize coordinated development of economy and environmental protection. Nevertheless, it is far from enough to realize win-win development of economy and environment only by relying on the government's administrative promotion and regulation. It is necessary to establish improved market mechanisms, especially the incentive mechanism.

In terms of environmental governance, efforts should be made to promote development of the carbon emissions trading market, release resources and environment pricing signals, improve resource efficiency, establish and improve the emission rights pricing mechanism and trading system on that basis, and explore the effective market-based operation of various pollution rights trading. In guiding and promoting development of ecological industries, efforts should be made to create a market environment conducive to scientific and technological innovation and suitable for demand for ecological products, develop green finance, and establish ecological product trading mechanism and markets, in an attempt to realize and expand the market value of ecological products.

6. CONCLUSION

In the new era, there are more and higher requirements for ecological conservation in China. Chinese President Xi Jinping pointed out at the National Conference on Ecological and Environmental Protection in May 2018 that China's in a crucial phase of ecological conservation in which we must carry forward despite heavy strain and immense pressure, a decisive stage in which we will supply more high-quality ecological goods to meet the growing demands of the people for a pristine environment, and also a period of opportunity in which we have the conditions and abilities necessary to resolve prominent environmental issues. This requires us to summarize previous work achievements, further deepen the environmental governance system reform, and strengthen the construction of pollution prevention and control technology and environmental governance capacity. By the middle of the twenty-first century, China will reach the objective of building a beautiful country in all-round improvements in the material, political, intellectual, social and ecological domains. At that time, environmentally friendly ways of living and developing will be fully formed, humans and nature will coexist in harmony, and modernization of our national governance systems and capacity in the environmental field will be fully realized.

Note

1 Yang Weimin: Topics about Ecological Civilization System Reform are discussed at 20 of the 38 Meetings of the Central Leading Group on Comprehensive Reform, People.cn, October 23, 2017.

www.ingramcontent.com/pod-product-compliance
Lightning Source LLC
LaVergne TN
LVHW021602060925
820435LV00004B/55